Lecture Notes in Mathematics

A collection of informal reports and seminars
Edited by A. Dold, Heidelberg and B. Eckmann, Zürich

Series: Forschungsinstitut für Mathematik, ETH, Zürich · Adviser: K. Chandrasekharan

98

Maurice Heins
University of Illinois, Urbana, Illinois

Hardy Classes on Riemann Surfaces

1969

Springer-Verlag Berlin · Heidelberg · New York

Library of Congress Catalog Card Number 75- 84833 · Title No. 3704

Contents

I wish to take this occasion to express my thanks to Professor Eckmann for his kind invitation to prepare these notes for the Springer Lecture Note Series and the secretarial help he has put at my disposal for this purpose. Part of the material of these notes - specifically that drawn from my papers [17] and [18] - was presented in a seminar held at the Forschungsinstitut für Mathematik of the E.T.H. June and July 1966 during one of the periods in which I enjoyed the very generous hospitality of the Forschungsinstitut für Mathematik. I wish to acknowledge gratefully the cogent comments made by members of the seminar, especially Professors Ch. Blatter, A.Huber, R.Narasimhan, and A.Pfluger. Since that time the program of the notes has expanded. An interim version was delivered during the summer of 1967 at the Conference on Complex Function Theory held at the University of Montreal. The notes which follow were prepared the summer of 1968 during the tenure of Grant GP 7405 of the National Science Foundation.

These notes treat the following topics which have been of continuing interest to me: (1) The role of the strongly subharmonic functions [10], first results concerning which were given by Solomentsev [34] in a euclidean setting - results which relate to Szegö's maximal theorem [36] and afford a natural frame of reference for aspects of the theory of Hardy classes. (2) The classification of Riemann surfaces and the theory of Hardy classes. (3) The role of the M.Riesz decomposition. (4) A theory of Hardy classes of vector-valued analytic functions.

One final observation. I have tried to keep the exposition reasonably self-contained and I hope that the usefulness of these notes will be thereby enhanced.

Chapter I

General Observations and Preliminaries

1. Some remarks concerning the theory of Hardy classes. The notion of a Hardy class was originally introduced in the following manner. Let $0 < p < +\infty$. A function f analytic on the open unit disk $\Delta = \{|z| < 1\}$ is said to belong to the Hardy class $H_p(\Delta)$ provided that

$$\int_0^{2\pi} |f(re^{i\theta})|^p d\theta = O(1), \quad 0 < r < 1. \tag{1.1}$$

By definition the Hardy class $H_\infty(\Delta)$ is the class of functions analytic on Δ and bounded. The study of p th means of functions analytic on Δ was initiated in Hardy's classical paper of 1915 [13]. Since that time the subject of Hardy classes has received very extensive treatment in both the classical setting of the open unit disk and, more recently, the setting of Riemann surfaces. Hardy classes are structurally rich and afford interesting examples of Banach spaces $(1 \leq p \leq +\infty)$ whose investigation remains a topic of current lively interest. For their study from this point of view particular reference is made to the monograph of Hoffman [21]. The first systematic study of Hardy classes on Riemann surfaces was given by Parreau in his thesis [26], in which many important notions were introduced. The present notes treat selected topics from the theory of Hardy classes on Rieman surfaces which have been of interest to me recently. There is no question of a systematic or exhaustive study of the subject.

During the decade following the appearance of Hardy's paper cited above fundamental results in the theory of Hardy classes were obtained. We are easily persuaded that this is the case when we call to mind (1) the celebrated paper of F. and M.Riesz [31] presented at the 1916 Scandinavian Congress of Mathematicians which treats in addition to other questions the boundary properties of functions belonging to the Hardy class $H_1(\Delta)$, (2) the maximal theorem of G. Szegö [36] given by him for the case $p = 2$ and subsequently by F. Riesz [31] for unrestricted positive p and (3) M.Riesz's theorem on the conjugate series of the Fourier series of a function belonging to $L_p[0,2\Pi]$, $1 < p < +\infty$, which admits direct interpretation and application in the theory of Hardy classes [32]. [It is to be observed in passing that the methods of the paper of Szegö and the subsequent paper of F. Riesz are quite different. The paper of Szegö appeals to the theory of Toeplitz forms.]

It was in this era that the fundamental papers [31] of F. Riesz on the theory of subharmonic functions appeared. The reference is apposite. Indeed, the p th power $(0 < p < +\infty)$ of the modulus of an analytic function is subharmonic. The condition (1.1) is equivalent to the condition that $|f|^p$ admits a harmonic majorant as we see with the aid of very elementary properties of subharmonic functions. We shall use this motivating fact to serve as a basis for the definition of Hardy classes on Riemann surfaces, though-to be sure - one may paraphrase the classical definition (1.1) through the mediation of mean values introduced in terms of "reasonable" subregions and the solution of associated Dirichlet problems. cf. p.35 of [26].

We return to the three results on Hardy classes which we cited. It is to be noted that the treatment given these questions by the Riesz brothers has a pronounced function-theoretic aspect. Thus in F. Riesz's treatment of Szegö's maximal principle for general p there enter the Parseval identity for power series, the notion of a Blaschke product, and the existence of analytic powers of a function analytic on a simply-connected region and free zeros.

Part of the argument given by M. Riesz for his conjugate series theorem is based on the Cauchy theory. Subsequently, an elegant proof of the M. Riesz theorem was given by P. Stein [35], which on examination is seen to exploit the subharmonicity (easily checked) of an auxiliary function introduced into the argument. We mention these facts because we shall see that not only will the theory of subharmonic functions be a very useful instrument of investigation in our study but also the basic results of the paper of F. and M. Riesz will appear as simple corollaries of a theorem concerning subharmonic functions subject to conditions considerably more general than those to which their counterparts in the theory of Hardy classes are subject. The subsuming theorem, given by Solomentsev [34] for the case of the unit ball, will be treated in Chapter II. We shall see that it also subsumes the Szegö maximal principle and is, in fact, appropriately termed the "Theorem of Szegö-Solomentsev". The general version for Riemann surfaces was given by me in [18].

It should be remarked that numerous treatments of the results of the Riesz bothers have been given. Of special interest from the potential-theoretic point of view are the paper of Gårding and Hörmander [10] and the paper of J.L. Doob [7].

2. Basic definitions and preliminaries. We shall be concerned with Riemann surfaces in the sense of Weyl-Radó and shall take for granted the usual elementary definitions and results, for which the reader is referred to one of the standard texts on Riemann surface theory, e.g. [1], [27]. In particular, we shall assume that the reader is familiar with such terms as local uniformizer, analytic, meromorphic, harmonic, subharmonic, superharmonic, taken in the context of Riemann surface theory.

We recall that the classical Harnack inequality for non-negative harmonic functions on the open unit disk has a qualitative counterpart for non-negative harmonic functions on a Riemann surface. Indeed, let S be a Riemann surface, let $a \in S$, and let Q denote the family of positive harmonic functions on S normalized to take the value 1 at a. We introduce λ, the lower envelope of Q, and μ, the upper envelope of Q. It is easily concluded with the aid of the cited classical Harnack inequality and the connectedness of S that λ is strictly positive and continuous and that μ is finite-valued and continuous. It is now obvious that we have the following qualitative form of the Harnack inequality holding for S:

_If_ u _is a non-negative harmonic function on_ S, _then_

$$u(a)\lambda \leqslant u \leqslant u(a)\mu . \tag{2.1}$$

The Harnack convergence theorem for monotone sequences of harmonic functions on a Riemann surface is a consequence of (2.1).

It is desirable to recall the notion of a Perron family and its utility in the study of harmonic majorants of subharmonic functions. For convenience we suppose, as we may, that the uniformizers α of S defining its conformal structure all have the open unit disk Δ as their domain. Let u be subharmonic on S (possibly the constant $-\infty$). Let α be such an allowed uniformizer and let $0 < r < 1$. Then by the (α,r) - Poisson modification v of u is meant the function defined on S by the condition that in $S - \alpha[\Delta(0;r)]$ it agree with u, while for points of $\alpha[\Delta(0;r)]$ it be given by

$$v[\alpha(rz)] = \frac{1}{2\pi}\int_0^{2\pi} u[\alpha(re^{i\theta})]k(\theta,z)\,d\theta\ , \quad |z| < 1\ .$$

Here $k(\theta,z)$ is $\mathrm{Re}[(e^{i\theta}+z)/(e^{i\theta}-z)]$, the Poisson kernel for the open unit disk, and $\Delta(0;r) = |z| < r$. [In general, we shall denote the open circular disk in $\mathbf{C}$ with center a and radius ϱ by $\Delta(a;\varrho)$.] It is standard that v is subharmonic. Let Φ be a family of functions subharmonic on S. We term Φ a Perron family provided that (1) $\max u,v \in \Phi$ whenever $u,v \in \Phi$, and (2) given an allowed uniformizer α and $u \in \Phi$, then the (α,r) - Poisson modification of u is also a member of Φ for each r satisfying $0 < r < 1$. It is to be observed that other equally convenient definitions for the notion of a Perron family are available. A fundamental fact concerning Perron families is given by the following trichotomy theorem:

The upper envelope of a Perron family is one of the following: the constant $-\infty$, the constant $+\infty$, a function harmonic on S.

A family ψ of functions subharmonic on S generates a Perron family $\Phi(\psi)$. More precisely, the intersection of all Perron families containing ψ is itself a Perron family containing ψ. It is this minimal family that we mean by the Perron family $\Phi(\psi)$ generated by ψ.

If ψ contains a member other than the constant $-\infty$ and, in addition, there exists a function h harmonic on S satisfying $u \leqslant h$, $u \in \psi$, then the set of subharmonic functions v on S satisfying $v \leqslant h$ is a Perron family containing ψ and the upper envelope of $\Phi(\psi)$ is majorized by h and is harmonic. We are led to the conclusion that there is a least harmonic function on S

which majorizes each member of ψ . It is in fact the upper envelope of $\Phi(\psi)$. We term it the <u>least harmonic majorant</u> of ψ . When ψ has a sole member u , we speak instead of the <u>least harmonic majorant</u> of u which we denote by Mu . The results of this paragraph may be recast dually for families of superharmonic functions. We obtain correspondingly the notion of the <u>greatest harmonic minorant</u> of a family of superharmonic functions as well as that of a given superharmonic function v (not the constant $+\infty$) . For such a v having a harmonic minorant we denote its greatest harmonic minorant by mv .

Suppose now that h is harmonic on S and that s_1 and s_2 are subharmonic on S , neither being the constant $-\infty$, and that $s_2 = h + s_1$. Then if one of the s_k has a harmonic majorant, so does the other and

$$Ms_2 = h + Ms_1 . \tag{2.2}$$

The first part of the assertion is routine to verify. To establish (2.2) we observe that $Ms_2 \geqslant h + s_1$ and hence $Ms_2 \geqslant h + Ms_1$. Similarly $s_2 \leqslant h + Ms_1$ and hence $Ms_2 \leqslant h + Ms_1$. The equality (2.2) follows.

The result that we have just proved admits application to the following classical lemma concerning differences of non-negative harmonic functions.

<u>Lemma 1</u> : <u>If</u> h <u>is the difference of non-negative harmonic functions on</u> S , <u>then there exists a unique</u> (p_1,p_2) , <u>where</u> p_1 <u>and</u> p_2 <u>are non-negative harmonic on</u> S , <u>satisfying</u> (1) $h = p_1 - p_2$, (2) <u>if</u> q_1 <u>and</u> q_2 <u>are non-negative harmonic on</u> S <u>and</u> $h = q_1 - q_2$, <u>then</u> $p_k \leqslant q_k$, $k = 1,2$.

Proof: We note that h^+ and $-h^-$ have harmonic majorants. From $h^+ = h + (-h^-)$ we conclude that $h = M(h^+) - M(-h^-)$. Since $h^+ \leqslant q_1$ and $-h^- \leqslant q_2$, we see that $M(h^+) \leqslant q_1$ and $M(-h^-) \leqslant q_2$. The lemma follows.

We now recall some fundamental concepts concerning non-negative harmonic functions which were introduced by Parreau in his thesis [26] . Let h be a non-negative harmonic function on S. We term h quasi-bounded provided that there exists a non-decreasing sequence of non-negative bounded harmonic functions on S which has limit h. (An equivalent definition is obtained when the restriction that the members of the sequence be non-negative is dropped.) We term h singular provided that the only non-negative bounded harmonic function on S majorized by h is the constant zero. Parreau showed that each non-negative harmonic function h on S admits a unique representation of the form

$$q + s \tag{2.3}$$

where q is quasi-bounded and s is singular. This result may be established very simply.

We first show uniqueness. To that end suppose that

$$q_1 + s_1 = q_2 + s_2 ,$$

where the q_k are quasi-bounded and the s_k are singular, $k = 1,2$. Let (b_n) be a non-decreasing sequence of non-negative bounded harmonic functions on S which has limit q_1. Then $(b_n - q_2)^+ \leqslant s_2, b_n$, the term on the left being subharmonic. Since $M[(b_n-q_2)^+]$ is bounded, non-negative, and is majorized by s_2, we conclude that $b_n \leqslant q_2$. On taking the limit we see that $q_1 \leqslant q_2$. By symmetry we conclude that $q_1 = q_2$. The uniqueness follows.

Existence. This question is easily treated. We introduce for each whole number n the greatest harmonic minorant, b_n, of $\min\{h,n\}$. The sequence (b_n) is non-decreasing and has as limit a quasi-bounded harmonic function, q, which is majorized by h. We assert that $h-q$ is singular. Suppose that b is a non-negative bounded harmonic function on S majorized by $h-q$ and suppose that m is a whole number satisfying $m \geqslant \sup b(S)$. Then $b_n + b \leqslant \min\{h,m+n\}$. Hence $b_n + b \leqslant b_{n+m}$. On letting $n \to \infty$ we conclude that $b \leqslant 0$, and consequently, b, being non-negative, is the constant zero. We conclude that $h-q$ is singular. The existence of a representation of the form (2.3) for h is established.

We term q of (2.3) the *quasi-bounded component* of h and s of (2.3) the *singular component* of h .

Sums of quasi-bounded (resp. singular) non-negative harmonic functions. We now show that *a convergent sum of quasi-bounded non-negative harmonic functions on* S *is quasi-bounded* and that the corresponding statement holds with "*singular*" replacing "*quasi-bounded*". Let q_k be a quasi-bounded non-negative harmonic function on S , $k = 0,1,\ldots$ and suppose that $\sum_0^\infty q_k$ is convergent. Since each q_k is representable as a convergent sum of bounded non-negative harmonic functions on S , the same is true for $\sum_0^\infty q_k$, which is consequently quasi-bounded. To treat the corresponding result for the case of singular non-negative harmonic functions on S we proceed as follows. With s_k singular replacing the above q_k we introduce b , a non-negative harmonic function on S , satisfying

$$b \leq \sum_0^\infty s_k ,$$

and observe that

$$(b-\sum_1^\infty s_k)^+ \leq \min\{b,s_o\},$$

whence we conclude that the least harmonic majorant of the left side of this inequality is zero. Hence

$$b \leq \sum_1^\infty s_k .$$

Proceeding inductively we see that this inequality holds with n , an arbitrary whole number, replacing 1 . Consequently $b = 0$ and hence $\sum_0^\infty s_k$ is singular. It is obvious that corresponding results hold as well for finite sums of quasi-bounded (resp. singular) non-negative harmonic functions.

Suppose now that h and H are non-negative harmonic functions on S satisfying $h \leq H$. Then h is quasi-bounded (resp. singular) when H is . It suffices to consider the canonical decompositions (2.3) for h , $H-h$ and H .

The Hardy classes $H_p(S)$, $1 \leq p < +\infty$. By definition $H_\infty(S)$ is just the set of bounded analytic functions on S . Given p, $0 < p < +\infty$, by $H_p(S)$ we understand the set of functions f analytic on S for which the subharmonic function $|f|^p$ has a harmonic majorant. It is routine that $H_\infty(S)$ is a complex Banach space when addition and multiplication by a scalar are introduced in the standard pointwise manner and the norm of a member f is taken to be $\sup |f|$. A more interesting question from a technical point of view is that of endowing $H_p(S)$ with a Banach space structure when $1 \leq p < +\infty$. This question is treated by Parreau in his thesis in terms of mean-values with the aid of the Minkowski inequality. However, as we shall now see, is it possible to approach the question internally without reference to exhaustions of S . We take advantage of the following elementary lemma.

Lemma 2: *Let* u *and* v *be non-negative superharmonic functions on* S *and let* $1 \leq p < +\infty$. *Then*

$$w = (u^{1/p} + v^{1/p})^p$$

is superharmonic on S .

Proof: It suffices to consider only the case where u and v are both harmonic (which will be the sole case considered in the application of the Lemma) and to treat the problem on a region of $\mathbb{C}$ by differential considerations. Further we put aside the trivial cases where either factor is zero or $p = 1$. Starting with

$$w^{1/p} = u^{1/p} + v^{1/p} ,$$

we obtain by differentiation (with respect to z)

$$w^{(1/p)-1} w_z = u^{(1/p)-1} u_z + v^{(1/p)-1} v_z , \tag{2.4}$$

and by a second differentiation (with respect to $\bar{z}$)

$$(\frac{1}{p} - 1) w^{(1/p)-2} |w_z|^2 + w^{1/p-1} w_{z\bar{z}}$$

$$= (\frac{1}{p} - 1)[u^{(1/p)-2}|u_z|^2 + v^{(1/p)-2}|v_z|^2] . \tag{2.5}$$

Writing (2.4) as

$$w^{\frac{1}{2p}}(w^{\frac{1}{2p}-1} w_z) = u^{\frac{1}{2p}}(u^{\frac{1}{2p}-1} u_z) + v^{\frac{1}{2p}}(v^{\frac{1}{2p}-1} v_z)$$

and applying the Cauchy-Schwarz-Buniakowsky inequality, we obtain

$$w^{\frac{1}{p}-2}|w_z|^2 \leqslant u^{\frac{1}{p}-2}|u_z|^2 + v^{\frac{1}{p}-2}|v_z|^2 ,$$

and applying this inequality to (2.5) we see that

$$w_{z\bar{z}} \leqslant 0 .$$

Hence w is superharmonic. The lemma follows.

Given $f \in H_p(S)$, $1 \leqslant p < +\infty$, we define h_f as $M(|f|^p)$. If g also belongs to $H_p(S)$, then

$$h_f^{1/p} + h_g^{1/p} \geqslant |f| + |g| \geqslant |f + g| .$$

From

$$(h_f^{1/p} + h_g^{1/p})^p \geqslant |f + g|^p$$

and the superharmonicity of the left side we conclude that $f + g \in H_p(S)$ and that

$$h_f^{1/p} + h_g^{1/p} \geqslant h_{f+g}{}^{1/p}. \tag{2.6}$$

It is immediate that $H_p(S)$, $0 < p < +\infty$, is a vector space over C when the standard pointwise definition of addition and multiplication by scalars is used as we see with the aid of the inequality

$$(a + b)^p \leqslant 2^p(a^p + b^p) ,$$

a and b being non-negative real numbers. But we have obtained much more in (2.6) which yields a triangle inequality. Given $q \in S$ as the (q-)norm of f we propose following Parreau

$$\|f\| = [h_f(q)]^{1/p}. \tag{2.7}$$

As noted, (2.6) yields the triangle inequality. The remaining norm conditions, i.e. (i) $\|f\| = 0$ if and only if $f = 0$ and (ii) $\|cf\| = |c|\,\|f\|$, $c \in C$, are routine to verify. Thanks to the qualitative Harnack inequalities (2.1) we see that changing of the reference point q yields an equivalent norm and that $H_p(S)$ is a Banach space in the sense of each norm when this is the case for some norm.

There remains to be shown that $H_p(S)$ is complete in the sense of the norm (2.7). Suppose that (f_n) is a Cauchy sequence in the sense of this norm. Let

$$u_{m,n} = M[(f_m - f_n)^p].$$

From

$$|f_m - f_n|^p \leqslant u_{m,n},$$

the fact that $u_{m,n}$ is small at q when m and n are large, and the right Harnack inequality of (2.1), we see that (f_n) is uniformly Cauchy on each compact subset of S. We proceed to show that f, the pointwise limit of (f_n), is a member of $H_p(S)$ and, in fact, is the limit of (f_n) in the sense of the norm topology. To that end, let η be a positive number and let ν be a positive whole number such that $u_{m,n}(q) \leqslant \eta$ for $m,n \geqslant \nu$. We see that for each $n > \nu$ we may select an increasing sequence of whole numbers, say μ, such that $(u_{\mu(m),n})$ tends to a non-negative harmonic function v_n on S as $m \to \infty$. We see at once that

$$|f - f_n|^p \leqslant v_n, \quad n \geqslant \nu,$$

whence we conclude that $f - f_\nu \in H_p(S)$ and hence that $f \in H_p(S)$. Further

$$\|f - f_n\|^p \leqslant v_n(q) \leqslant \eta,$$

for $n \geqslant \nu$. Given the arbitrariness of η, the completeness of $H_p(S)$ is seen to follow.

Chapter II

The Theorem of Szegö - Solomentsev

1. The de la Vallée Poussin condition. We shall be concerned with a continuous non-decreasing function φ with domain $\{-\infty \leq x < +\infty\}$ which takes non-negative real values and satisfies the following two conditions: (a) the restriction of φ to the real line $\mathbb{R}$ is convex, (b) $\lim_{+\infty} x^{-1}\varphi(x) = +\infty$. It will be convenient to introduce ψ, the inverse of the restriction of φ to $\{\xi \leq x < +\infty\}$, where $\xi = \max\{x : \varphi(x) = \varphi(-\infty)\}$. Clearly $\varphi \circ \psi$ is the identity map on the domain of ψ and $\psi \circ \varphi(x) = \max\{\xi, x\}, -\infty \leq x < +\infty$. Further ψ is concave on its domain when $\xi > -\infty$ and the restriction of ψ to $\{\varphi(-\infty) < x < +\infty\}$ is concave when $\xi = -\infty$. Functions of the same kind as φ (without the convexity requirement) were introduced by de la Vallée Poussin for formulating a necessary and sufficient condition for uniform integrability and were subsequently employed for this purpose also by Nagumo. cf. [11]. For the use of the de la Vallée Poussin condition in the study of harmonic and subharmonic functions we cite the work of Doob [6] and Yamashita's paper [38] depending on the work of Doob. A proof of Theorem 1 below is given in Yamashita's cited paper. The proof to be given here, which is very different, proceeds by "internal" arguments.

<u>Theorem</u> 1: <u>Let</u> u <u>be a quasi-bounded non-negative harmonic function on</u> S. <u>Then there exists an allowed</u> φ <u>such that</u> $\varphi \circ u$ <u>possesses a harmonic majorant</u>.

Proof: Putting aside the trivial case where u is bounded we start by intoducing auxiliary families of functions depending on a parameter taking real values greater than inf u. Given λ, $\inf u < \lambda < +\infty$, we introduce $\Gamma_\lambda = \{u(q) = \lambda\}$ and $\Omega_\lambda = \{u(q) > \lambda\}$. We note that the family of functions s subharmonic on Ω_λ and satisfying: (1) $s \leq u|\Omega_\lambda$, (2) $\lim\sup_q s \leq 0$, $q \in \Gamma_\lambda$, is a Perron family containing the constant zero. The definition of a Perron family given for the case of a Riemann surface is to be used with the obvious changes for the case of an open subset $(\neq \emptyset)$ of a Riemann surface. The upper envelope of this family majorizes $(u-\lambda)|\Omega_\lambda$. It will now be shown that the upper envelope has limit zero at each point of Γ_λ. To that end, we consider the interior, ω, of a small "semicircle" whose frontier consists of a "diameter" which is a Jordan arc lying on Γ_λ and containing a given point of Γ_λ as a non-endpoint, and of a "semicircumference" lying in Ω_λ save for its endpoints. [The case where u may have a stationary point on Γ_λ is to be taken into regard.] We introduce the bounded harmonic function on ω which vanishes continuously at each point of the "diameter" less its endpoints and has limit $u(q)$ at each point q of the "semicircumference" not an endpoint. It majorizes each $s|\omega$. The asserted boundary behavior of the upper envelope of the Perron family in question is seen to follow.

We define s_λ as the function with domain S whose restriction to Ω_λ is just the upper envelope in question and whose restriction to $S - \Omega_\lambda$ is the constant zero. We see that s_λ is subharmonic and $s_\lambda < u$. The family $\{s_\lambda\}$ is non-increasing and the family $\{Ms_\lambda\}$ is also. It will be seen that $\lim_{\lambda\to+\infty} Ms_\lambda = 0$, thanks to the quasi-boundedness of u. This fact will play a fundamental part in constructing an allowed φ having the property stated in Theorem 1.

We define the auxiliary function t_λ in a manner similar to that used in defining s_λ with $S - \bar{\Omega}_\lambda$ replacing Ω_λ. Thus we take $t_\lambda|(S - \bar{\Omega}_\lambda)$ as the upper envelope of the family of functions t subharmonic on $S - \bar{\Omega}_\lambda$ and satisfying: (1) $t \leqslant u|(S - \bar{\Omega}_\lambda)$, (2) $\limsup_q t \leqslant 0$, $q \in \Gamma_\lambda$. We take $t_\lambda|\bar{\Omega}_\lambda$ as the constant zero. We see that also t_λ is subharmonic and has u as a harmonic majorant. The families $\{t_\lambda\}$ and $\{Mt_\lambda\}$ are non-decreasing.

The function v_λ is defined as the function satisfying: (1) $v_\lambda|\bar{\Omega}_\lambda = u|\bar{\Omega}_\lambda$, (2) $v_\lambda|(S - \bar{\Omega}_\lambda)$ is the least positive harmonic function with domain $S - \bar{\Omega}_\lambda$ having limit λ at each point of Γ_λ. The existence of such a harmonic function on $S - \bar{\Omega}_\lambda$ is readily concluded with the aid of Perron methods. The function v_λ is superharmonic. The following equality

$$u = t_\lambda + v_\lambda \tag{1.1}$$

holds. It suffices to check that the two sides of (1.1) take the same value at each point $q \in S \cdot \bar{\Omega}_\lambda$, the control for points of $\bar{\Omega}_\lambda$ being a trivial consequence of the definitions of t_λ and v_λ. Now on the one hand, we see using property (2) of v_λ that $u(q) - t_\lambda(q) \geq v_\lambda(q)$. On the other hand, $0 \leq u(q) - v_\lambda(q) \leq u(q)$ and $u - v_\lambda$ vanishes on Γ_λ. Hence $u(q) - v_\lambda(q) \leq t_\lambda(q)$. The asserted equality follows and (1.1) is thereby established.

Thanks to (1.1) we conclude that the family $\{v_\lambda\}$ is non-increasing. We show that $\lim_{\lambda\to+\infty} v_\lambda$, which is non-negative harmonic on S, is actually the constant zero. From this fact we conclude, using the inequality $v_\lambda \geq s_\lambda$, that $\lim_{\lambda\to+\infty} Ms_\lambda = 0$. At all events, $\lim v_\lambda$, being majorized by u, is quasi-bounded. Suppose that b is a non-negative harmonic function on S bounded above by 1 which is majorized by $\lim v_\lambda$. Then for points $q \in S - \bar{\Omega}_\lambda$ we have

$$\lambda b(q) \leq u(q), \tag{1.2}$$

whence we conclude that $b = 0$. The inequality (1.2) may be established as follows. First, we note that $b|(S - \bar{\Omega}_\lambda)$ is the least non-negative harmonic function on $S - \bar{\Omega}_\lambda$ having the limit $b(q)$ at each point $q \in \Gamma_\lambda$. Otherwise there would exist a non-negative harmonic function on $S - \bar{\Omega}_\lambda$, not the constant zero, majorized by $v|(S - \bar{\Omega}_\lambda)$ and vanishing continuously on Γ_λ. This would violate the minimality property of $v|(S - \bar{\Omega}_\lambda)$. To avoid unnecessary complications we suppose for the moment that there are no stationary points of u on Γ_λ. We note that $b|(S - \bar{\Omega}_\lambda)$ is the limit of a sequence (b_n), where b_n is the least non-negative harmonic function on $S - \bar{\Omega}_\lambda$ which has the limit $b(q)$ at each point q of the union $\Gamma_\lambda^{(n)}$, of a finite number of relatively compact open subarcs of Γ_λ, and $\Gamma_\lambda^{(n)} \uparrow \Gamma_\lambda$. Since $\lambda b_n(q) \leq v_\lambda(q), q \in S - \bar{\Omega}_\lambda$, we conclude that $\lambda b(q) \leq u(q)$ for the q in question. The restriction on λ is now dropped by an obvious limit argument. The inequality (1.2) is thereby established. Now $b = 0$ as a consequence of (1.2). We conclude that $\lim v_\lambda$ is <u>singular</u>, and thereupon, since it is also quasi-bounded, that it is the constant zero.

We observe that for each allowed λ the quotient Ms_λ/u has the property that "it tends to 1 as $u(q)$ tends to $+\infty$". More precisely given $\alpha < 1$, there exists a positive number μ such that

$$\frac{Ms_\lambda(q)}{u(q)} > \alpha$$

for $q \in \Omega_\mu$. To see this, we use the inequality $u - \lambda \leq s_\lambda$ and conclude that

$$\frac{Ms_\lambda(q)}{u(q)} > 1 - \frac{\lambda}{\mu}$$

for $q \in \Omega_\mu$. The asserted limit behavior of Ms_λ/u follows.

Since $\lim_{\lambda\to+\infty} Ms_\lambda = 0$, there exists an increasing sequence (λ_k), $\lim \lambda_k = +\infty$, such that $\Sigma M(s_{\lambda_k})$ is convergent. The sum is, of course, positive harmonic on S. Using the property of Ms_λ established in the preceding paragragh, we conclude that with

$$f(x) = \inf \{\Sigma Ms_{\lambda_k}(q) \; : \; q \in \Gamma_x\}, \; \inf u < x < +\infty,$$

we have

$$\lim_{x\to+\infty} x^{-1} f(x) = +\infty. \tag{1.3}$$

We are now in a position to construct an admitted φ such that $\varphi \circ u$ has a harmonic majorant. To that end, we introduce the set

$$\{(x,y) \; : \; x > \inf u, \; y \geq f(x)\} \tag{1.4}$$

and thereupon its convex hull K. We next define $\theta(x)$ as $\inf \{y : (x,y) \in K\}$, $\inf u < x < +\infty$. It is immediate that θ is convex. Because of (1.3) the set (1.4) lies above some line with a given positive slope and hence $\lim_{x\to+\infty} x^{-1}\theta(x) = +\infty$. When θ is non-decreasing, we define φ by $\varphi(x) = \theta(x)$, $\inf u < x$, and $\varphi(x) = \inf \theta$, $x \leq \inf u$. Otherwise θ has a minimum, say at μ, and we define $\varphi(x) = \theta(x)$, $x \geq \mu$, $\varphi(x) = \theta(\mu)$, $x < \mu$. In either case φ so constructed is allowed.

The proof of Theorem 1 follows on noting that the subharmonic function $\varphi \circ u$ is majorized by the harmonic function ΣMs_{λ_k}.

2. The theorem of Szegö-Solomentsev. We now turn to a converse question and consider the consequences of the hypothesis thst $\varphi \circ u$ has a harmonic majorant on S where u is subharmonic on S and φ is allowed in the sense of §1, this Ch. The fundamental result in this direction is the theorem of Szegö type first given by Solomentsev [34] and subsequently by Gårding and Hörmander [10]. Extensions of the work of Solomentsev to regions of euclidean space having a reasonably regular character were given by Privalov and Kuznetsov. The work of Solomentsev was preceded by work of Privalov which considered the special case: $\varphi(x) = (x^+)^\alpha$, $1 < \alpha < +\infty$. [I am indebted to Professor Lars Gårding for these bibliographical indications.] For Riemann surfaces the condition of this paragraph appears in the suggestive paper of R.Nevanlinna [25] and in Parreau's thesis [26], in which latter work the special case that u is the modulus of a harmonic function is considered. The work of Parreau in this direction finds its impetus in the paper of R. Nevanlinna just cited. In what follows we establish a theorem of Szegö-Solomentsev type for the case of Riemann surfaces [Theorem 2 below] given by us in [18], together with additional consequences of the hypothesis [Theorem 3 below] yielding maximality properties not given in our paper. The proof of Theorem 2 will be based on a lemma which is a special case of Theorem 14 of Parreau's thesis; in the lemma [Lemma 1 below] u is taken to be non-negative harmonic. The proof of the lemma is simple and somewhat more immediate than the proof given by Parreau of Theorem 14 of his thesis. The principal theorem of this section states

Theorem 2: Suppose that $\varphi \circ u$ has a harmonic majorant. Then u^+ and $\varphi \circ u^+$ have harmonic majorants and, in fact, Mu^+ and $M(\varphi \circ u^+)$ are quasi-bounded.

If, in addition, u is not the constant $-\infty$, then u admits a unique representation of the form

$$Q - s - g, \tag{2.1}$$

where Q is the difference of quasi-bounded non-negative harmonic functions on S, s is a singular, non-negative harmonic function on S, and g is a non-negative super-

harmonic function on S *satisfying* $mg = 0$; *and further not only does* $\varphi \circ Q$ *have a harmonic majorant but also*

$$M(\varphi \circ Q) = M(\varphi \circ Mu) = M(\varphi \circ u). \tag{2.2}$$

There is a companion maximal theorem. It states

Theorem 3: Under the hypothesis of Theorem 2 *the upper envelope*, H, *of the family*, Φ, *of subharmonic functions* v *on* S *satisfying*

$$M\varphi \circ v = M\varphi \circ u$$

is a member of Φ. *When* H *is not the constant* $-\infty$, *it is the difference of quasi-bounded non-negative harmonic functions on* S. *When* ξ (of §1, this Ch.) *is not* $-\infty$, *then* Φ *is exactly the set of functions* v *subharmonic on* S *which satisfy*

$$M \max \{\xi, v\} = H.$$

When $\xi = -\infty$ *but* u *is not the constant* $-\infty$, Φ *is the set of subharmonic functions on* S *of the form* (2.1) *with* $Q = H$.

We start with the following lemma.

Lemma 1 (Parreau): *Under the hypothesis of* Theorem 2 *if* u *is non-negative harmonic on* S, *then* u *and* $M\varphi \circ u$ *are quasi-bounded*.

Proof: Given a positive number c, there exists a positive number d such that

$$cx \leqslant \varphi(x) + d$$

for all real x. With s denoting the singular component of u, we obtain from

$$cu \leqslant \varphi \circ u + d \leqslant M\varphi \circ u + d,$$

on comparing the singular components of the extreme members, the inequality

$$cs \leqslant M\varphi \circ u.$$

Given the arbitrariness of c, we see that $s = 0$ and hence that u is quasi-bounded.

We now introduce (b_n), a non-decreasing sequence of bounded non-negative harmonic functions on S with limit u. From

$$\varphi \circ u \leqslant \lim M\varphi \circ b_n \leqslant M\varphi \circ u,$$

and the fact that the middle term is quasi-bounded, we see that

$$M\varphi \circ u = \lim M\varphi \circ b_n$$

and that $M\varphi \circ u$ is quasi-bounded.

We now turn to the proof of Theorem 2. It is obvious that

$$\varphi(x^+) \leqslant \varphi(x) + \varphi(o),$$

φ being non-decreasing and taking non-negative values. We conclude that $\varphi \circ u^+$ has a harmonic majorant. From

$$u^+ \leqslant \psi \circ M(\varphi \circ u^+)$$

and the fact that the right side is superharmonic, we see that u^+ has a harmonic majorant and that

$$\varphi \circ (Mu^+) \leqslant M(\varphi \circ u^+). \tag{2.3}$$

Hence by Lemma 1 we conclude that Mu^+ is quasi-bounded. From (2.3) we infer, using the fact that φ is non-decreasing, that

$$M[\varphi \circ (Mu^+)] = M(\varphi \circ u^+).$$

Hence by Lemma 1 we see that $M(\varphi \circ u^+)$ is quasi-bounded.

We continue, supposing for the remainder of the proof that u is <u>not</u> the constant $-\infty$. Since u has a quasi-bounded non-negative harmonic majorant, we see, using the canonical representation (2.3), Ch.1, of a non-negative harmonic function, that u admits a representation of the asserted form (2.1). It is to be observed that

$$u = Mu^+ - m[(Mu^+) - u] - g$$

where g is a superharmonic function on S satisfying $mg = 0$. [We remark in passing that the g in question are just the Green's potentials on S generated by non-negative mass distributions. cf. F. Riesz's theorem on the representation of superharmonic functions possessing a harmonic minorant.] The uniqueness of the representation (2.1) follows at once from the fact that a non-negative superharmonic function on S, not the constant $+\infty$, admits a unique representation of the form

$$q + s + g, \tag{2.4}$$

where q is a quasi-bounded non-negative harmonic function on S, s is a singular non-negative harmonic function on S, and g is a non-negative superharmonic function on S satisfying $mg = 0$. The uniqueness of the representation (2.4) follows when we consider two such representations for a given allowed superharmonic function, note that the sum of the "q" and "s" terms of each representation is just the greatest harmonic minorant of the given superharmonic functions, and thereupon invoke (2.3),Ch.I.

There remains to be shown that $\varphi \circ Q$ has a harmonic majorant and that the equality (2.2) holds. We suppose that $Q = q_1 - q_2$ where q_1 and q_2 are quasi-bounded non-negative harmonic functions on S. We introduce a positive number ε. Let c be a positive number satisfying

$$c + \psi[\varphi(-\infty) + \varepsilon] > 0.$$

Starting with the observation

$$u \leq \psi \circ (M\varphi \circ u + \varepsilon),$$

we conclude on taking the least harmonic majorant of the left side that

$$Q - s \leq \psi \circ (M\varphi \circ u + \varepsilon),$$

from which inequality we obtain

$$c + q_1 \leq s + q_2 + [c + m\psi \circ (M\varphi \circ u + \varepsilon)].$$

On noting that the left side of this inequality is majorized by the quasi-bounded component of the right side, we conclude that

$$Q \leqslant m\psi \circ (M\varphi \circ u + \varepsilon),$$

whence it follows that

$$M\varphi \circ Q \leqslant M\varphi \circ u + \varepsilon.$$

Given the arbitrariness of ε we conclude the equality (2.2). It is to be observed that the role of c is ancillary. Once it has served its purpose, to permit comparison of non-negative harmonic functions, it no longer appears in the argument.

The proof of Theorem 2 is complete.

Proof of Theorem 3: We first consider the case where $\xi = -\infty$ and u is the constant $-\infty$. Here H is the constant $-\infty$ and so is trivially a member of Φ. The remaining assertions of the theorem are vacuously true in the present case. We put this case aside.

We are now assured that $\psi \circ [M(\varphi \circ u)]$ is superharmonic. If $v \in \Phi$, we have

$$v \leqslant \max\{\xi, v\} \leqslant \psi \circ [M(\varphi \circ u)]. \tag{2.5}$$

We assert that

$$H = m\psi \circ [M(\varphi \circ u)] \in \Phi, \tag{2.6}$$

To see this we note that

$$\varphi \circ [m\psi \circ (M\varphi \circ u)] \leqslant M\varphi \circ u,$$

which yields

$$M\varphi \circ [m\psi \circ (M\varphi \circ u)] \leqslant M\varphi \circ u. \tag{2.7}$$

Using (2.5) we obtain

$$M\varphi \circ u \leqslant M\varphi \circ [m\psi \circ (M\varphi \circ u)].$$

On combining this inequality with (2.7) we see that $m\psi \circ (M\varphi \circ u) \in \Phi$. On taking the greatest harmonic minorant of the right side of (2.5) we see that $m\psi \circ (M\varphi \circ u) \geq v$, $v \in \Phi$. The assertation (2.6) follows. The first statement of Theorem 3 has been established in all cases.

We turn to the second assertion. We assume that H is not the constant $-\infty$. Suppose that u is the constant $-\infty$. Then $\xi > -\infty$, H is the constant taking the value ξ, and trivially H is the difference of quasi-bounded non-negative harmonic functions on S. Suppose that u is not the constant $-\infty$. We introduce Q of (2.1) relative to u. By (2.2) of Theorem 2 we see that $Q \leq H$. By the first paragraph of Theorem 2 we see that H is majorized by a quasi-bounded non-negative harmonic function on S. It follows readily now that H is the difference of quasi-bounded non-negative harmonic functions on S.

The developments which now follow are preparatory to the proofs of the last two assertions of Theorem 3. We fix $v \in \Phi$ and let α satisfy $\xi < \alpha < +\infty$. We introduce Φ_α, the set of w subharmonic on S which satisfy

$$M\varphi \circ w = M\varphi \circ \max\{\alpha, v\}.$$

It is to be noted that $\varphi \circ \max\{\alpha, v\}$ does indeed have a harmonic majorant as we see with the aid of the inequality $\varphi \circ \max\{\alpha, v\} \leq \varphi(\alpha) + \varphi \circ v$. We let H_α denote the upper envelope of the family Φ_α. Using the representation given by (2.6) for H we see that $H \leq H_\alpha$ and that H_α is non-decreasing in α. Further the convexity of φ yields

$$\varphi \circ H_\alpha - \varphi \circ (M \max\{\alpha, v\})$$

$$\geq \varphi'_+(\alpha)(H_\alpha - M \max\{\alpha, v\}), \qquad (2.8)$$

$\varphi'_+(\alpha)$ being the right derivative of φ at α. From this inequality we conclude that

$$0 = M\varphi \circ H_\alpha - M\varphi \circ (M \max\{\alpha, v\})$$

$$\geq \varphi'_+(\alpha)(H_\alpha - M \max\{\alpha, v\}),$$

and since $\varphi'_+(\alpha) > 0$, we obtain the formula

$$H_\alpha = M \max\{\alpha, v\}. \tag{2.9}$$

Suppose now that $\xi > -\infty$. Using the obvious inequality

$$\max\{\xi, v\} \leqslant \max\{\alpha, v\} \leqslant \max\{\xi, v\} + (\alpha - \xi),$$

we conclude that

$$H \leqslant \lim_{\alpha \to \xi} M \max\{\alpha, v\} = M \max\{\xi, v\}.$$

In the opposite direction, using (2.5) and (2.6) we obtain

$$M \max\{\xi, v\} \leqslant H.$$

Thus we find that for each $v \in \Phi$ we have the equality

$$M \max\{\xi, v\} = H. \tag{2.10}$$

Suppose that v is subharmonic on S and satisfies (2.10). Since $\varphi \circ v = \varphi \circ \max\{\xi, v\}$, we see that $M\varphi \circ H$ is a harmonic majorant of $\varphi \circ v$. We conclude from

$$M\varphi \circ v = M\varphi \circ M \max\{\xi, v\}$$

that $v \in \Phi$. Hence Φ is precisely the set of functions v subharmonic on S and satisfying (2.10). The third assertion of Theorem 3 is thereby established.

There remains to be considered the final assertion of the theorem. Here, we recall, $\xi = -\infty$ and u is not the constant $-\infty$. The notations Φ_α and H_α are to be taken in the senses to be <u>specified presently</u>. We take $\alpha \in R$, let $\varphi_\alpha = \max\{\varphi(\alpha), \varphi\}$, and note that

$$\varphi_\alpha \leqslant \varphi + [\varphi(\alpha) - \varphi(-\infty)].$$

We now let Φ_α denote the class of w subharmonic on S which satisfy

$$M\varphi_\alpha \circ w = M\varphi_\alpha \circ v$$

where v is a given member of Φ. We now let H_α denote the upper envelope of Φ_α. We let Q now denote the term in question in the representation of the form (2.1) of v. Using the third assertion of the present theorem and the fact that

$$M\varphi_\alpha \circ Q = M\varphi_\alpha \circ v,$$

which is a consequence of Th. 2, this Ch., we conclude that

$$H_\alpha = M \max\{\alpha, Q\}.$$

Since

$$m\psi \circ (M\varphi \circ v) \leqslant m\psi \circ (M\varphi_\alpha \circ v),$$

and $Q \in \Phi$, we conclude that

$$Q \leqslant H \leqslant H_\alpha.$$

We write $Q = q_1 - q_2$, where q_1 and q_2 are quasi-bounded non-negative harmonic functions on S, and introduce (b_n), a non-decreasing sequence of non-negative bounded harmonic functions on S with limit q_2. For a given n we have

$$q_1 - b_n \geqslant \max\{\alpha, Q\}$$

when α is large and negative. Hence for such α the inequality

$$q_1 - b_n \geqslant H_\alpha \geqslant H$$

holds. We conclude that $Q \geqslant H$ and thereupon that $Q = H$. We see that the Q associated with each member of Φ is H.

Suppose that v is subharmonic on S and of the form (2.1) and that the associated Q is equal to H. From $\varphi \circ v \leqslant \varphi \circ H$ we see that $\varphi \circ v$ has a harmonic majorant. It is now immediate that

$$M\varphi \circ v = M\varphi \circ Q = M\varphi \circ H.$$

We conclude that in the case where $\xi = -\infty$ and u is not the constant $-\infty$ the family Φ consists exactly of the functions of the form (2.1) with $Q = H$.

The proof of Theorem 3 is complete.

The following corollary of Theorem 3 permits us to relate H to an arbitrary member of Φ when $\xi > -\infty$.

Corollary: Given that $\xi > -\infty$. For each $v \in \Phi$ not the constant $-\infty$, we have

$$H = q + \xi,$$

where q is defined as follows. With Q taken relative to v in the sense of (2.1) we let $q = M[(Q-\xi)^+]$.

Proof: It suffices to note that

$$M \max\{\xi, Q\} = \xi + M[(Q-\xi)^+].$$

[cf. L. 1, Ch.I. for the significance of q.]

3. A comparison theorem. We suppose that φ_1 satisfies the conditions imposed on φ in §1, this Ch., that u is subharmonic on S, that $\varphi \circ u$ has a harmonic majorant, and that Φ retains the meaning assigned to it in the preceding section.

Theorem 4: If there exists $v \in \Phi$ such that $\varphi_1 \circ v$ has a harmonic majorant, then $\varphi_1 \circ u$ has a harmonic majorant.

Proof: We put aside the trivial case where either u or v is the constant $-\infty$. The case where $\xi = \max\{x : \varphi(x) = \varphi(-\infty)\} = -\infty$ is an immediate consequence of Theorems 2 and 3. Indeed, with Φ_1 being taken relative to φ_1 and v as Φ is to φ and u, we see that in this case $\Phi \subset \Phi_1$. We put this case aside.

To continue, let $Q(v)$ denote the term Q of (2.1) taken relative to v. We have

$$u \leqslant M \max\{\xi, Q(v)\} \tag{3.1}$$

by Theorem 3. Using the representation (2.1) for v we obtain

$$\max\{\xi, Q(v)\} \leqslant \max\{\xi^+, v\} + s + g$$

and conclude that

$$M \max\{\xi, Q(v)\} \leq M \max\{\xi^+, v\}.$$

Since

$$\varphi_1 \circ \max\{\xi^+, v\} \leq \varphi_1 \circ v + \varphi_1(\xi^+),$$

$\varphi_1 \circ \max\{\xi^+, v\}$ has a harmonic majorant. Hence so does $\varphi_1 \circ M \max\{\xi^+, v\}$ and by (3.1) we see that $\varphi_1 \circ u$ does. The theorem is established.

4. Remark. More generally, one may consider φ satisfying the conditions imposed in §1, this Ch., save in place of $\lim_{x\to+\infty} x^{-1}\varphi(x) = +\infty$ we require that φ be not constant. Here the role of Q is taken over by a function of the form $p-q$, where p and q are non-negative harmonic functions on S, q is quasi-bounded and $m \min\{p,q\} = 0$. However, we shall not have occasion to make use of this more general hypothesis and therefore leave the matter with the above summary indication.

5. The classical Theorem of F. and M.Riesz. The celebrated 1916 paper of F. and M.Riesz has many aspects, but a central result, which may be taken as a basis for proving the other principal results of the paper, is the fact that a function belonging to the class $H_1(\Delta)$ admits a Poisson - Lebesgue representation. From it one may conclude the mean convergence (order p) of the family $f_r : \theta \to f(re^{i\theta})$ as $r \to 1$ where $f \in H_p(\Delta)$ $\circ$ $p \geq 1$, a characterization of the Fatou boundary function of such an f, and the absolute continuity of a measure on the unit circumference which annihilates the "boundary functions" of functions continuous on $\bar{\Delta}$ and analytic in Δ.

Now the Poisson - Lebesgue representation in question is easy to establish with the aid of Th. 2, this Ch.. In fact, let $f \in H_1(\Delta)$. Let $u = \log|f|$ and let $\varphi = \exp$ extended to $R\cup\{-\infty\}$ by the stipulation that $\varphi(-\infty) = 0$. We conclude by Theorem 2 that $|f|$ has a quasi-bounded harmonic majorant, say q. From

$$-q \leq \mathrm{Re} f \leq q,$$

we see that Ref is the difference of quasi-bounded non-negative harmonic functions on the unit disk; the same is true for $\mathrm{Im} f$. The fact that f admits a Poisson - Lebesgue representation follows when we observe that the quasi-bounded non-negative harmonic functions on Δ are precisely the harmonic functions on Δ given by the Poisson-Lebesgue integrals with non-negative integrands.

6. Szegö's Maximal Theorem. This theorem was given originally by Szegö [36] in the H_2 setting with the aid of the theory of Toeplitz forms. Subsequently, F.Riesz [31] gave the extended form of this theorem, which we shall consider immediately.

We suppose that p is finite and positive and that $f \in H_p(\Delta)$ but f is not the constant 0. Let f^* denote the Fatou radial limit function of f. [It is to be recalled that such an f admits a representation as the quotient of bounded analytic functions, as we see readily with the aid of Theorem 2 applied to $u = \log|f|$, $\varphi(x) = \exp(px)$, $x \in R$, $\varphi(-\infty) = 0$, and that thanks to the Fatou limit theorem for bounded analytic functions on Δ and the theorem of F. and M. Riesz (also from the 1916 paper) which asserts that the Fatou radial limit function of a bounded analytic function on Δ, not the constant zero, takes the value zero at most on a set of measure zero, such a quotient of bounded analytic functions on Δ possesses Fatou radial limits p.p.. The facts are, of course, standard.] The Szegö theorem may be formulated as follows:

<u>Theorem</u> 5: (a) $\log|f^*(e^{i\theta})|$ <u>and</u> $|f^*(e^{i\theta})|^p$ <u>are integrable on</u> $[0,2\Pi]$. (b) <u>Given</u> $F > 0$ <u>on</u> $[0,2\Pi]$ <u>such that</u> $\log F$ <u>and</u> F^p <u>are integrable. If</u> G <u>is analytic on</u> Δ <u>and satisfies</u>

$$\log|G(z)| = \frac{1}{2\Pi}\int_0^{2\Pi} \log F(\theta)k(\theta,z)d\theta, |z| < 1, \qquad (6.1)$$

<u>then</u> $G \in H_p(\Delta)$ <u>and</u> $|G^*(e^{i\theta})| = F(\theta)$ p.p.. (c) <u>Let</u> h <u>be such a</u> G <u>with</u> $F(\theta) = \log|f^*(e^{i\theta})|$. <u>Then</u>

$$|f| \leqslant |h|. \qquad (6.2)$$

That is, h is maximal in modulus in the family of functions in $H_p(\Delta)$ for which the modulus of the Fatou radial limit function agrees p.p. with $|f^*|$. The set of such maximal functions is just the set of functions of the form ηh where η is a constant of modulus one. (d) We have

$$(M|f|^p)(z) = \frac{1}{2\Pi}\int_0^{2\Pi} |f^*(e^{i\theta})|^p k(\theta,z)\,d\theta, |z|<1. \tag{6.3}$$

Our main concern is to identify (c) of Theorem 5 as a precursor of Theorem 2. We take u of Theorem 2 as $\log|f|$ and $\varphi(x) = \exp(px), x \in R, \varphi(-\infty) = 0$. Now (c) implies that u so chosen is majorized by a quasi-bounded non-negative harmonic function on Δ. We are led to a representation of the form (2.1) for u using classical facts concerning analytic functions having a given harmonic function as the logarithm of the modulus and the representation of a bounded analytic function on Δ. Using the known radial limit behavior of a Blaschke product (resp. of a singular non-negative harmonic function on Δ) we see that, in fact, $\log|h|$ is the term Q of the representation (2.1) of u.

We now show that (c) and (d) may be subsumed under Theorem 2. The representation (2.1) is available and g is the negative of the logarithm of the modulus of a Blaschke product. We conclude the integrability of $\log|f^*(e^{i\theta})|$. The term Q of (2.1) is here given by (6.1) with $F(\theta) = |f^*(e^{i\theta})|$. We conclude that $|f| \leq |h|$. Since $\varphi \circ Q$ has a harmonic majorant, $h \in H_p(\Delta)$. The assertion (c) follows. Now $M|f|^p$ is quasi-bounded by Theorem 2 and the meanvalue on $C(0;r)$ of $M|f|^p - |f|^p$, which is non-negative tends to 0 as $r \to 1$,— we are using the reasoning of Gårding and Hörmander [10]. We conclude that the representation (6.3) holds.

We remark that (b) of Theorem 5 generalizes: If U and $\varphi \circ U$ are integrable on $[0,2\Pi]$, then

$$u(z) = \frac{1}{2\Pi}\int_0^{2\Pi} U(e^{i\theta})k(\theta,z)\,d\theta, |z|<1,$$

is such that $\varphi \circ u$ has a harmonic majorant (Jensen inequality) and $(\varphi \circ u)^* = \varphi\circ U$ p.p.

The following remarks are appropriate. If f of Theorem 5 also satisfies the condition that $|f^*| \in L_q[0,2\Pi]$, $0 < p < q < +\infty$, then $f \in H_q(\Delta)$. This result is well-known. It may be proved by appeal to (c) and (b) of Theorem 5. In terms of the general developments of §2,3, this Ch., we may proceed thus. With $\varphi(x) = \exp(px)$, $\varphi_1(x) = \exp(qx)$, $\varphi(-\infty) = \varphi_1(-\infty) = 0$, $u = \log|f|$, the term Q of the representation (2.1) of $\log|f|$ is $\log|h|$ and $|h|^q = \varphi_1 \circ (\log|h|)$ has a harmonic majorant and hence by Theorem 4 we see that $f \in H_q(\Delta)$. To be sure, the use of Theorem 4 is hardly called for in the obvious situation lying at hand. However it is to be noted that in Theorem 4 the intervening mediating function need not be the upper envelope of Φ as it is here.

7. An application of the extended Theorem of Szegö-Solomentsev. It is well-known that the Hardy class $H_2(\Delta)$ may be brought to bear on the study of bounded analytic functions on Δ, in particular, in connection with the theory of Toeplitz forms. cf. [12]. In fact, if f is a complex-valued function on Δ such that $\theta_f : g \to fg$ maps the closed unit ball of $H_2(\Delta)$ into itsself, then f is an analytic function on Δ of modulus at most one. Further such a map θ_f is norm preserving if and only if the quasibounded component of $m \log(1/|f|) = 0$,or equivalently if and only if $M(|f|)^2 = 1$. In this section we shall consider a theorem concerning the so-called PL functions which implies the result just quoted as well as a corresponding one for $H_p(S)$, $0 < p < +\infty$. Related theorems will be met in Chs. IV and V of these notes.

A function on S is termed a PL function provided that it admits a representation of the form $\exp \circ u$ where u is subharmonic on S. As throughout we understand that $\exp(-\infty)$ is 0. A PL function is subharmonic. Since the modulus of an analytic function is PL, it is to be anticipated that results concerning PL functions will have significance for the theory of analytic functions. By way of general reference we cite the tract of Radó on subharmonic functions [29].

We fix $a \in S$. For each PL function G possessing a harmonic majorant we define $\nu(G)$ as the value at a of MG. We let F be a map of S into the set of non-negative reals and introduce $\Theta_F(G) = FG$, G a PL function. X is to denote a subset of $\{\nu(G) \leq 1\}$ containing a member not the constant 0. We show

Theorem 6: (a) If F is continuous on S and $\Theta_F(X) \subset X$, then F is PL and satisfies $F \leq 1$. (b) An F such that Θ_F maps the set of PL functions G having harmonic majorants into itself has the property that

$$\nu[\Theta_F(G)] = \nu(G) \tag{7.1}$$

for all such G if and only if F is PL, $F \leq 1$, and the quasi-bounded component of $m \log(1/F)$ is 0, or equivalently, if and only if F is PL and $MF = 1$. (c) If F is PL, $F \leq 1$, and (7.1) holds for one allowed G, not the constant zero, then (7.1) holds for all allowed G.

Proof: (a) Let G be a member of X, not the constant 0. By induction we see that $F^n G \in X$ for each non-negative whole number n. Let $b \in S$, $G(b) \neq 0$. By the Harnack inequality we have

$$M(F^n G)(b) = O(1),$$

whence we conclude that $F(b) \leq 1$. Since the set of admitted b is dense on S and F is continuous, we conclude that $F \leq 1$. Since $F^n G$ is PL for each positive whole number n, we see that

$$\log F + \frac{1}{n} \log G$$

is subharmonic on S for each such n. We conclude by a limit argument, using the continuity of F, that F is a PL function. The assertion (a) is not valid if the stipulation of continuity is dropped, however a valid assertion is obtained if the requirement of continuity is replaced by the requirement that F be a PL function, the first half of the conclusion being redundant.

(b) F is PL and has a harmonic majorant when Θ_F maps the set of PL functions having harmonic majorants into itself since the constant 1 belongs to the set. If Θ_F satisfies (7.1) for all allowed G, then $F \leq 1$. Taking G as the constant 1 in (7.1) we see that

$$(MF)(a) = 1$$

and hence by the maximum principle for harmonic functions, $MF = 1$. By Th. 3, this Ch., we see that since $M1 = 1$, the quasi-bounded component of $m \log(1/F)$ is the constant zero.

To proceed in the opposite direction, given that F is PL, $F \leq 1$, and the quasi-bounded component of $m \log(1/F)$ is the constant zero, we see that the term Q in the representation (2.1), this Ch., of $\log F$ is the constant zero. Hence the corresponding terms in the representations of $\log G$ and $\log(FG)$ are equal, and consequently $M(FG) = MG$, when we consider G not the constant zero. The equality is trivially true when G is the constant zero. The equality so obtained implies (7.1). [We see incidentally that the condition (7.1) implies for an allowed F that $M(FG) = MG$, all G in question.] Returning to the first sentence of this paragraph we see from the fact that the term Q relative to $\log F$ is the constant zero that $MF = 1$, thanks to Th. 2, this Ch., (b) is established.

(c) This follows on using the maximum principle for harmonic functions. For if G_o has the stated property, $M(FG_o) = M(G_o)$ and consequently the term Q relative to $\log F$ is the constant zero. Hence the quasi-bounded component of $m \log(1/F)$ is zero and the assertion follows from (b).

We have noted in the first paragraph of this section the equivalence of the conditions, the quasi-bounded component of $m \log(1/|f|)$ is 0, $M(|f|^2) = 1$, for analytic functions on Δ of modulus at most one. This equivalence is a special case of the following property of PL functions F not exceeding one: If the quasi-bounded component of $m \log(1/F)$ is 0, then $M(F^p) = 1$, $0 < p < +\infty$. If $M(F^p) = 1$ for some positive real p, then the quasi-bounded component of $m \log(1/F)$ is zero. This observation follows at once with the aid of Theorem 3, this Ch..

Application to analytic functions. We turn to the study of maps of the type considered at the beginning of this section. We start afresh, defining θ_f now in a manner appropriate for our present purposes.

Let $0 < p < +\infty$. Let $f, g : S \to C$ and define $\theta_f(g)$ to be fg. We suppose that Y is a subset of the "closed unit ball" (relative to a) of $H_p(S)$, containing a member not the constant zero. [We are allowing all positive p. The closed unit ball (relative to a) is construed as the set

$$\{g \in H_p(S),\ \nu(|g|^p) \leq 1\}.]$$

We have the following easy consequence of Theorem 6.

Theorem 7: (a) If $\theta_f(Y) \subset Y$, then there exists a unique analytic function on S of modulus at most one, say f_1, such that f differs from f_1 on a set clustering at no point of S. (b) θ_f is a map of $H_p(S)$ into itself satisfying

$$\nu[|\theta_f(g)|^p] = \nu(|g|^p), \tag{7.2}$$

$g \in H_p(S)$, i.e. θ_f is a "norm"-preserving map of $H_p(S)$ into itself if and only if f is an analytic function on S of modulus at most 1 for which the quasi-bounded component of $m \log(1/|f|)$ is zero. (c) If f is analytic on S and of modulus at most one and if (7.2) holds for some $g \in H_p(S)$, not the constant zero, then (7.2) holds for all $g \in H_p(S)$.

Proof: (a) It suffices to take $F = |f|^p$, $X = \{|g|^p : g \in Y\}$ and to observe that from the analyticity of g and fg for some $g \in Y$, not the constant zero, f is analytic at each point of the complement of a part of S which clusters at no point of S.

(b) On taking g as the constant 1 we see that f is analytic on S. An application of (a) of the present theorem with Y the closed unit ball (relative to a) shows that if θ_f is "norm"-preserving, then f is of modulus at most one. Consequently when θ_f is "norm"-preserving, F satisfies the hypotheses of (c) of Theorem 6. Hence (7.1) holds for all allowed G and, a fortiori, (7.2) holds for all $g \in H_p(S)$. Hence the quasi-bounded component of $m \log(1/|f|) = 0$. The converse follows on applying (b) of Theorem 6 to $F = |f|^p$.

(c) This is immediate when (c) of Theorem 6 is applied to F.

We remark that θ_f is a "norm"-preserving map of $H_p(S)$ onto itself if and only if f is a constant of modulus 1. The "if" is trivial. When θ_f maps $H_p(S)$ onto itself and is "norm"-preserving, the constant 1 belongs to $\theta_f[H_p(S)]$ and consequently f does not take the value 0 anywhere, and further $\theta_{1/f}$ maps $H_p(S)$ onto itself and is "norm"-preserving. We conclude that $|f| = 1$ and hence that f is a constant of modulus one.

Chapter III

A Classification Problem for Riemann Surfaces

1. Statement of problem and results. The following question was proposed to me by Mr. C. W. Neville in 1967: If for some $p, 0 < p < +\infty$, the class $H_p(S)$ has a non-constant member, does there exist a non-constant bounded analytic function on S? We shall see that it is not the case that the answer is affirmative for all S and shall, in addition, obtain a chain of strict inclusion relations supplementing the known inclusion relations given in the standard accounts of the classification theory of Riemann surfaces. cf. Ch.IV of [1].

It will be convenient to introduce the following notations. By O_p we shall understand the set of Riemann surfaces S for which $H_p(S)$ contains only constant members. Here and throughout this chapter $0 < p < +\infty$. Since "$\subset$" is always taken to mean weak inclusion in the course of these notes and we shall have frequent occasion to deal with strict inclusion in this chapter, we shall employ "$\subsetneq$" to mean strict inclusion. Given $p < q < +\infty$, from

$$x^p < x^q + 1, \quad 0 \leq x < +\infty,$$

we conclude that $H_p(S) \supset H_q(S)$ and hence $O_p \subset O_q$. We introduce

$$\underline{O}_p = \cup_{0<q<p} O_q$$

and

$$\bar{O}_p = \cap_{p<q} O_q.$$

By O_{BA} we shall understand the set of Riemann surfaces on which there do not exist non-constant bounded analytic functions. Clearly $O_p \subset O_{BA}$.

We shall also consider a "null" class which will be seen to appear at the lower end of our classifying chain. We recall that a non-constant meromorphic function f on a hyperbolic Riemann surface S is termed <u>Lindelöfian</u> provided that for all $a \in S$ we have

$$\Sigma_{f(s)=w}\, n(s;w)\, g_s(a) < +\infty, \quad w \neq f(a),$$

where $n(s;f)$ is the multiplicity of f at s and g_s is Green's function for S with pole s [15]. The stipulation that a non-constant meromorphic function on S be Lindelöfian is equivalent to the requirement that it have bounded Nevanlinna characteristic. In special case of a non-constant analytic function f the stipulation that f be Lindelöfian is equivalent to the condition that $\overset{+}{\log}|f|$ have a harmonic majorant. Given S hyperbolic by $LA(S)$ we understand the set of analytic functions on S which either are constant or Lindelöfian. It is obvious from the inequality

$$\overset{+}{\log}\, x \leqslant x^p/p, \quad x \geqslant 0,$$

that $LA(S) \supset H_p(S)$. We define O_{LA} as the set of Riemann surfaces S which are either parabolic or else are hyperbolic and such that $LA(S)$ consists exactly of the complex constants on S. The inclusion, $O_{LA} \subset O_p$, holds.

The principal result developed in this chapter is simply the following chain of strict inclusions, the first of which was established by Parreau in his thesis [26, p. 90]:

$$\left.\begin{aligned} &O_{LA} \prec \cap_{0<q<+\infty} O_q \prec \underline{O}_p \prec O_p \prec \bar{O}_p \\ &\prec \cup_{0<q<+\infty} O_q \prec O_{BA}. \end{aligned}\right\} \qquad (1.1)$$

We observe that the last inclusion asserts the existence of a Riemann surface S on which there exists no non-constant bounded analytic function but which nevertheless is such that for each p there exists a non-constant member of $H_p(S)$. Our construction (§3) furnishes such an S for which there exists a non-constant analytic function on S belonging to

$$\cap_{0<p<+\infty} H_p(S).$$

Indeed, the construction of §3 shows that given φ satisfying the condition of §1 of Ch.II, regardless of its rate of growth there exists $S \in O_{BA}$ such that there exists f analytic on S, not constant, and such that $\varphi \circ |f|$ has a harmonic majorant. Whether one can refine the last inclusion of (1.1) by introducing null classes O_{φ} (the set of Riemann surfaces S such that when f is analytic on S and $\varphi \circ |f|$ has a harmonic majorant, f is constant) with very rapidly growing φ is not settled.

Not much appears to be known about the counterpart of (1.1) in the more restrictive planar theory. We shall conclude this chapter with some simple remarks concerning this question.

2. $O_{LA} < \cap_{0<q<+\infty} O_q$. We give an account of Parreau's result which is based directly on P. J. Myrberg's original example [24] of a Riemann surface belonging to O_{BA} but on which there exists a non-constant bounded harmonic function. It is no exaggeration to say that the classification theory of Riemann surfaces owes its impetus to Myrberg's example which was startling when it was published. The role of Myrberg's idea in the constructions of the subsequent sections of this chapter will be apparent.

We consider the analytic structure $\mathfrak{G}$ (analytische Gebilde in the sense of Weyl, for which cf. [19], [37]) annihilating

$$w^2 - \cos \frac{\Pi}{2}z = 0.$$

It is a parabolic Riemann surface of infinite genus with one Kerékjártó boundary element. The center map "from $\mathfrak{G}$ into the z-plane" has constant valence two and has ramification points exactly at the preimages of the odd integers, each such point having ramification order equal to one. The analytic structure in question is termed "transcendental hyperelliptic" for obvious reasons. The preimage of $\{|z| \leq 1/2\}$ with respect to the center map has two components, each a homeomorph of the closed unit disk. The desired Riemann surface S is obtained by removing one of these components from $\mathfrak{G}$. It is a Riemann surface of the type considered by P. J. Myrberg and has the property that it belongs to O_{BA} but does admit non-constant bounded harmonic functions and, of course, is hyperbolic. This last fact together with the bounded valence of the center map shows that there exists on S a Lindelöfian analytic function (namely, the restriction to S of the center map associated with $\mathfrak{G}$). Hence $S \notin O_{LA}$. Suppose now that $f \in H_p(S)$. We introduce for $1/2 < |z| < +\infty$, the function ψ given by

$$\psi(z) = [f(s) - f(t)]^2,$$

where $\{s,t\}$ is the preimage of z with respect to the center map. We note that ψ is analytic on its domain. From the inequality

$$|\varphi(z)|^{p/2} \leqslant 2^p[|f(s)|^p + |f(t)|^p]$$

and the fact that $|f|^p$ has a harmonic majorant h on S we conclude the inequality

$$|\psi(z)^{p/2} \leqslant 2^p[h(s) + h(t)];$$

the right-hand side of which represents a non-negative harmonic function on $\{1/2 < |z| < +\infty\}$. We conclude, using the local behavior of a positive harmonic function in a punctured disk and the Cauchy inequalities for the Laurent coefficients of ψ, that ψ has a removable singularity at ∞. Since ψ takes the value 0 at each odd integer we see that ψ vanishes identically. Let Π denote the restriction to S of the center map associated with $\mathfrak{G}$. We conclude that $f = \tilde{f} \circ \Pi$ where $\tilde{f}$ is analytic on C. For $1/2 < |z| < +\infty$ we have

$$|\tilde{f}(z)|^p \leqslant [h(s) + h(t)]/2.$$

where $\{s,t\}$, as above, is the preimage of z. We conclude, using the reasoning applied to ψ, that $\tilde{f}$ and, consequently, f are constant. Hence $S \in \cap_{0<q<+\infty} O_q$. The inclusion relation of Parreau is established.

Remark. The boundedness of ψ and $\tilde{f}$ near ∞ may be inferred with the aid of the theorem of Szegö-Solomentsev. This observation affords a second approach for showing the constancy of f.

3. $\cup_{0<q<+\infty} O_q \prec O_{BA}$. Given φ satisfying the conditions imposed in §1 of Ch.II, we show that with O_φ as defined in §1, this Ch.

$$O_\varphi \prec O_{BA}.$$

When

$$\lim_{x\to+\infty} \frac{\log \varphi(x)}{\log x} = +\infty,$$

we have

$$O_p \subset O_\varphi, \quad 0 < p < +\infty.$$

The strict inclusion of the first sentence of this paragraph follows. It is immediate that $O_\varphi \subset O_{BA}$. We show that the inclusion is strict.

To that end, we introduce E_1, a copy of $\Delta(0;2) - \{0\}$, and distinguish (slits along) the segments $[1/2^{2n+2}, 1/2^{2n+1}]$, $n = 0,1, \ldots$; E_2, a copy of $\Delta(0;2) - \{0\}$, and distinguish (slits along) the above segments and also (along) the segments

$$[\varepsilon_{2n}, \varepsilon_{2n+1}],$$

$n = 0,1, \ldots$, where (ε_n) is strictly increasing, $\varepsilon_0 > 1$ and $\lim \varepsilon_n = 2$. E_{3+n}, a copy of $\Delta(0;3+n)$ and distinguish (a slit along) $[\varepsilon_{2n}, \varepsilon_{2n+1}]$, the indices n running through the non-negative whole numbers. The ε_n will be further restricted in the course of the construction.

We construct the desired surface S by joining E_2 to E_1 along their common distinguished slits in the standard manner (i.e. the upper edge of such a slit of a given copy being joined to the lower edge of the corresponding slit of the other copy) and joining E_{3+n} to E_2 along their common distinguished slit also in the standard manner, $n = 0,1, \ldots$

Let us be specific in the case of the construction lying at hand. For the remaining constructions we shall content ourselves with the inexact descriptive language that we have just employed grosso modo. By the copy E_n we understand the image of $\Delta(0;2) - \{0\}$ with respect to the map $z \to (z,n)$, $n = 1,2$; by the copy E_{3+n} we understand the image of $\Delta(0;3+n)$ with respect to $z \to (z,3+n)$, $n = 0,1, \ldots$. By the "joining" of the preceding paragraph we understand that we are concerned with a 1-complex dimensional manifold S, to be presently described, endowed with a conformal structure rendering a "natural" projection map into $\mathbb{C}$ analytic. The underlying set of S is the union of

$$\{(z,1) : 0 < |z| < 2\};$$

$$\{(z,2) : 0 < |z| < 2; z \neq 2^{-(n+1)}, n=0,1, \ldots\};$$

$$\{(z,n+3) : |z| < n+3, z \neq \varepsilon_{2n}, \varepsilon_{2n+1}\},$$

$$n = 0,1, \ldots .$$

Given (a,k) of this set we define $\delta(a,k;r)$ for sufficiently small positive r as follows. (1) When a is not on a segment distinguished for E_k and r does not exceed the distance from a to the union of the segments distinguished for E_k, it is

$$\{(z,k) : |z - a| < r\}.$$

(2) When k is 1 or 2 and a is a point of a segment distinguished for E_1 but not an end point, and r is no greater than the minimum of the distances from a to the endpoints of the distinguished segment containing $a, \delta(a,k;r)$ is the union of

$$\{(z,k) : |z - a| < r, \operatorname{Im} z \geq 0\}$$

and

$$\{(z,3 - k) : |z - a| < r, \operatorname{Im} z < 0\}.$$

(3) When k is 1 and a is an endpoint of a segment distinguished for E_1, and r is at most the distance from a to the set of the remaining endpoints of the segments distinguished for $E_1, \delta(a,1;r)$ is the union of

$$\{(z,1) : |z - a| < r\}$$

and

$$\{(z,2) : 0 < |z - a| < r\}.$$

(4) When $k = 2$ or $3+n$ and $a \in [\varepsilon_{2n}, \varepsilon_{2n+1}]$, we proceed as we have just done in (2)

and (3), mutatis mutandis.

The topology on S is taken as that generated by the family of allowed $\delta(a,k;r)$, which is a base for the topology. This topology renders S a 1-complex dimensional manifold and the projection map π, taking each point of S into its first component, an interior map of S onto C. There is a conformal structure on S rendering π analytic and it is determined up to an equivalence. We suppose that S is endowed with such a conformal structure. [We may be specific here if we wish and consider as uniformizers the inverses of the $\pi|\delta(a,k;r)$ in cases (1), (2); the continuous maps θ mapping $\Delta(0;r^{1/2})$ onto $\delta(a,k;r)$ satisfying $\pi[\theta(z)] \equiv a + z^2$ in case (3); and the appropriate counterparts in case (4).]

We now show that $S \in O_{BA}$. To that end we consider a bounded analytic function on S, say f, and show that it is constant. Let Ω denote the component of $\Pi^{-1}(\Delta)$ consisting of points of E_1 and E_2. The map $\Pi|\Omega$ has image $\Delta - \{0\}$, at each point of which the valence is 2. Paraphrasing the Myrberg argument we introduce

$$F(z) = [f(s) - f(t)]^2, \quad 0 < |z| < 1,$$

where $\{s,t\}$ is the pre-image with respect to $\Pi|\Omega$ of z. The function F is bounded and analytic, and further vanishes at the points $2^{-(n+1)}, n = 0,1, \ldots$. We conclude that $f(s) = f(t)$. We thereupon introduce ψ_n, the inverse of the restriction of Π to E_n less the union of the segments distinguished for $E_n, n = 0,1, \ldots$. The functions $f \circ \psi_1$ and $f \circ \psi_2$ take the same values at each point of the domain of ψ_2. It follows that $f \circ \psi_{3+n}$ and $f \circ \psi_1$ take the same values at each point of, say, $\{|z| < 2, \mathrm{Im}\, z < 0\}$, $n = 0,1, \ldots$. It is now concluded that f takes the same value at each point of the preimage of a given point of C and thereupon that $f = \tilde{f} \circ$ where $\tilde{f}$ is analytic on C and bounded. By the theorem of Liouville $\tilde{f}$ is constant and hence so is f. Consequently $S \in O_{BA}$.

It remains to be shown that given φ there exists an allowed (ε_n) such that $S \notin O_\varphi$. For that purpose it suffices to show that (ε_n) may be so chosen that $\varphi \circ |\Pi|$ has a harmonic majorant. We proceed as follows. For each non-negative whole number ν we introduce an auxiliary surface S_ν formed by joining $E_2^{(\nu)}$ to E_1 and $E_3, \ldots, E_{3+\nu}$ to $E_2^{(\nu)}$ as above where $E_2^{(\nu)}$ differs from E_2 in that only the segments distinguished for E_1 and the segments $[\varepsilon_{2k}, \varepsilon_{2k+1}]$, $k = 0, \ldots, \nu$, are distinguished for $E_2^{(\nu)}$.

For a given non-negative whole number ν we let Π_ν denote the projection map associated with S_ν, defined as above. It is trivial from the construction of S_ν that Π_ν is bounded. Let h_ν denote $M\varphi \circ |\Pi_\nu|$. It will be shown that <u>with</u> $\varepsilon_0, \ldots, \varepsilon_{2\nu+2}$ <u>held fast</u>, $h_{\nu+1}$ <u>tends pointwise to</u> h_ν <u>on</u> $S_\nu - \{(\varepsilon_{2\nu+2}, 2)\}$ <u>as</u> $\varepsilon_{2\nu+3}$ <u>tends to</u> $\varepsilon_{2\nu+2}$. This fact will be of fundamental importance for the disposition of the parameters ε_n.

We fix r satisfying

$$0 < r < \min\{2 - \varepsilon_{2\nu+2}, \varepsilon_{2\nu+2} - \varepsilon_{2\nu+1}\}$$

and restrict $\varepsilon_{2\nu+3}$ to be less than $r + \varepsilon_{2\nu+2}$. We define C as the image with respect to $z \to (z,2)$ of

$$\{|z - \varepsilon_{2\nu+2}| = r\}.$$

We introduce Ω_1 and Ω_2, the components of $S_{\nu+1} - C$, the indices being specified by the requirement that $(\varepsilon_{2\nu+2}, 2) \in \Omega_2$. Let u be the largest non-negative harmonic function on Ω_1 which vanishes continuously on C and is majorized by

$$M\varphi \circ |\Pi_\nu|\Omega_1|.$$

Let v be the least positive harmonic function on Ω_1 taking the boundary value 1 continuously at each point of C. We introduce two functions with domain S_ν less the image of $[\varepsilon_{2\nu+2}, \varepsilon_{2\nu+3}]$ with respect to $z \to (z,2)$. The first, U, is defined to be the extension of u which takes the value 0 at each point of its domain not in Ω_1. The second, V, has the same domain as U, is an extension of v, and takes the value 1 at the points of its domain not in Ω_1. It is evident that U is subharmonic and V is superharmonic on their domain. It is readily seen that

$$U + \varphi(3 + \nu + 1)V$$

is superharmonic on its domain. Further for each point s of the common domain of U and V we have

$$U(s) \leq h_\nu(s), h_{\nu+1}(s) \leq U(s) + \varphi(3 + \nu + 1)V(s).$$

Consequently, if H with domain $S_\nu - \{(\varepsilon_{2\nu+2}, 2)\}$ is the pointwise limit of a sequence of $h_{\nu+1}$ (the parameter $\varepsilon_{2\nu+3}$ tending to $\varepsilon_{2\nu+2}$), we conclude that

$$0 \leq H(s) - h_\nu(s) \leq \varphi(3 + \nu + 1)V^*(s), \ s \in S_\nu - \{(\varepsilon_{2\nu+2}, 2)\},$$

where $V^*(s) = v(s)$, $s \in \Omega_1$, and $V^*(s) = 1$ elsewhere in $S_\nu - \{(\varepsilon_{2\nu+2}, 2)\}$, and thereupon the equality of $H(s)$ and $h_\nu(s)$ at each point of $S_\nu - \{(\varepsilon_{2\nu+2}, 2)\}$. It follows that $h_{\nu+1}$ tends pointwise to h_ν on $S_\nu - \{(\varepsilon_{2\nu+2}, 2)\}$ as $\varepsilon_{2\nu+3}$ tends to $\varepsilon_{2\nu+2}$ as asserted above.

The parameters ε_n are to be chosen as follows. We fix ε_n when n is 1 or even. We fix a point $a \in S_o$, whose second component is 1. We define (ε_{2n+3}) recursively to satisfy the requirement that

$$h_{\nu+1}(a) < 2^{-\nu} + h_\nu(a), \ \nu = 0, 1, \ldots .$$

The surface S_ν is now replaced by S_ν^*, the region obtained by removing from S_ν the image of

$$\bigcup_{\mu \geq \nu + 1} [\varepsilon_{2\mu}, \varepsilon_{2\mu + 1}]$$

with respect to $z \to (z, 2)$. The sequence (S_ν^*) is increasing and exhausts S. Since the value at a of $M\varphi \circ |\Pi| S_\nu^*|$ does not exceed $h_\nu(a)$, which is less than $2 + h_o(a)$, we conclude that $\varphi \circ |\Pi|$ has a harmonic majorant. Hence, since Π is not constant, $S \notin O_\varphi$.

4. $\underline{O}_p \prec O_p$. Since $\underline{O}_p \subset O_p$, it suffices to construct a Riemann surface $S \in O_p - \underline{O}_p$. The following observation is central in the construction:

_If_ $0 < \alpha < 1$, _then the identity map of_ $\mathfrak{G} = \{|\mathrm{Arg}\, z| < \alpha\Pi/2\}$ belongs to $H_1(\mathfrak{G})$.

The proof is an immediate consequence of the inequality

$$|z| < \frac{\mathrm{Re}z}{\cos\frac{\alpha\Pi}{2}} , \quad z \in \mathfrak{G}.$$

We note that the observation just made yields the result that a function f analytic on a Riemann surface S which has positive real part belongs to $H_\alpha(S), 0 < \alpha < 1$, given by Smirnov in the case $S = \Delta$. In fact, it suffices to note that $\theta \circ f \in H_1(S)$, where θ is the analytic α th power of the identity map of $\{\mathrm{Re}z > 0\}$ onto itself satisfying $\theta(1) = 1$.

To proceed with our construction, we introduce: E_1, a copy of $\Delta(0;2) - \{0\}$, for which the segments $[2^{-(2n+2)}, 2^{-(2n+1)}]$, $n = 0,1, \ldots$, have been distinguished; E_2, a copy of $\Delta(0;2) - \{0\}$ for which the above segments and also

$$[e^{2\Pi ik/m}, \tfrac{3}{2}e^{2\Pi ik/m}], \tag{4.1}$$

$k = 0, \ldots, m-1$, m being a given positive integer to be further restricted, are distinguished; E_{3+k}, $k = 0, \ldots, m-1$, given by

$$E_{3+k} = \{(e^{2\Pi ik/m}\exp(\frac{\mathrm{Log}\ z}{p}), z, k) : \mathrm{Re}z > 0\},$$

for which the "parameter" segment $[1,(3/2)^p]$ is distinguished. The underlying set of the surface to be constructed consists of the image of $\Delta(0;2) - \{0\}$ with respect to $z \to (z,1)$, the image of $\Delta(0;2) - \{0\}$ less $\{2^{-(n+1)}: n = 0,1, \ldots\}$ with respect to $z \to (z,2)$ and the E_{3+k} less the images of 1 and $(3/2)^p$ with respect to

$$z \to (e^{2\Pi ik/m} \exp(\frac{\mathrm{Log}\ z}{p}), z, k). \tag{4.2}$$

The surface S is obtained by joining E_2 to E_1 in the standard manner along the slits distinguished in common for E_1 and E_2 and by joining E_{3+k} to E_2 along (4.1), $k = 0, \ldots, m-1$. The precise technical meaning to be attributed to the joining of E_2 to E_1 should be clear from the developments of §3 - it consists in stating what $\delta(a;r)$ enter when a is of the form $(x,1)$ with x a point of one of the segments $[2^{-(2n+2)}, 2^{-(2n+1)}]$ or is of the form $(x,2)$ where x is now a point of such a segment but is not an endpoint. The joining of E_{3+k} to E_2 consists in discerning the $\delta(e^{2\Pi ik/m}x,2;r)$ where $1 \leqslant x \leqslant 3/2$ and the

$$\delta(e^{2\Pi ik/m}x^{1/p}, x,k;r) \tag{4.3}$$

for $1 < x < (3/2)^p$.

When $1 < x < 3/2$ we take for r small $\delta(e^{2\Pi ik/m}x,2;r)$ as the union of the image of

$$\{|z - x| < r, \text{Im}z \geqslant 0\} \tag{4.4}$$

with respect to

$$z \to (e^{2\Pi ik/m}z, 2) \tag{4.5}$$

and the image with respect to (4.2) of the preimage of

$$\{|z - x| < r, \text{Im}z < 0\}, \tag{4.6}$$

with respect to the restriction of

$$z \to \exp\left(\frac{\text{Log } z}{p}\right) \tag{4.7}$$

to a disk centered at x^p on which (4.7) is univalent. The corresponding definition for (4.3) is given on replacing 'x' by '$x^{1/p}$', '$\text{Im}z \geqslant 0$' by '$\text{Im}z < 0$' in (4.4), and '$\text{Im}z < 0$' by '$\text{Im}z \geqslant 0$' in (4.6). The set $\delta(e^{2\Pi ik/m}x,2;r)$ is defined for $x = 1$ or $3/2$ as the union of the image of $\Delta(x;r)$ with respect to (4.5) and the image with respect to (4.2) of the preimage of $\Delta(x;r) - \{x\}$ with respect to (4.7) as restricted. For other points of E_{3+k} we define

$$\delta(e^{2\Pi ik/m}\exp\left(\frac{\text{Log } w}{p}\right), w,k;r)$$

for small r as the image of $\Delta(w;r)$ with respect to (4.2).

The projection map Π on $\bigcup_{0}^{3+(m-1)} E_k$ is defined as the map taking each point of

$\overset{3+(m-1)}{\underset{0}{\cup}} E_k$ into its first component. For convenience after the following paragraph $\Pi|S$ will be denoted by Π. A conformal structure is introduced on S as in §3.

We fix m as the smallest integer exceeding $2p$. When $p < 1/2$, $m = 1$ and $\Pi(E_3) = C - \{0\}$. When $p \geq 1/2$, $\Pi(E_{3+k}) \cap \Pi(E_{4+k}) \neq \emptyset$, $k = 0, \ldots, m - 2$, and the union of the $\Pi(E_{3+k})$, $k = 0, \ldots, m - 1$, is $C - \{0\}$.

We first show that $S \in O_p$. Let $f \in H_p(S)$. We introduce Ω, the region of S consisting of the points of E_1 and E_2 whose first components are of modulus less than 1. Applying the argument of §2, this Ch. we see that $f|\Omega$ takes the same value at points having the same first component and further is bounded. Using an argument paralleling that used in §3, this Ch. we conclude that f admits a unique representation of the form $g \circ \Pi$ where g is an entire function. It now suffices to show that

$$g(z) = O(|z|),$$

z large, for then g, and consequently, f will be constant so that $S \in O_p$.

As an auxiliary step we consider the behavior at ∞ of a function $F \in H_p(D)$ where

$$D = \{\text{Re} z > 0, |z| > 2^p\}. \tag{4.8}$$

By the Szegö-Solomentsev theorem (Th.2, Ch.II) we see that $M|F|^p$ is quasi-bounded. Applying standard estimates to the Poisson integral for a half-plane we conclude that

$$|F(z)|^p = O(|z|) \tag{4.9}$$

for z large and satisfying $|\text{Arg } z| \leq \alpha\Pi/2$ where α is given satisfying $0 < \alpha < 1$. We observe that a Poisson integral representation is available for the restriction of $M|F|^p$ to $\{\text{Re} z > c\}$ where $c > 2^p$.

Applying (4.9) to f restricted to the image of D with respect to (4.2), $k = 0, \ldots, m - 1$, we conclude that

$$|g[e^{2\Pi ik/m}\exp(\frac{\text{Log } z}{p})]|^p = O(|z|)$$

for large z satisfying $|\text{Arg } z| \leq \alpha\Pi/2$, $k = 0, \ldots, m-1$. Taking α sufficiently near 1, we conclude that $|g(z)|^p = O(|z|^p)$ for z large and thereupon the constancy of g.

We now show that $\Pi \in H_q(S)$, $0 < q < p$, whence the asserted property of S will follow. For $k = 0, \ldots, m-1$ we let Ω_{3+k} denote the image of

$$\{\text{Re} z > 0, |z - (3/2)^p| > (3/2)^p - (1/2)^p\}$$

with respect to (4.2) and C_k the image of

$$\{|z - (3/2)^p| = (3/2)^p - (1/2)^p\}$$

with respect to the same map. By the observation of the first paragraph of this section

$$\Pi|\Omega_k \in H_q(\Omega_k),$$

$k = 3, \ldots, 2+m$, since $z \to \exp(\frac{q}{p} \text{ Log } z)$ maps the right-half plane onto the sector $\{\text{Arg } z < \frac{q}{p}\frac{\Pi}{2}\}$. We denote $\cup_{3 \leq k \leq 2+m}\Omega_k$ by Ω and $\cup_{3 \leq k \leq 2+m} C_k$ by C. We let u denote the function with domain S whose restriction to $S - \Omega$ is the constant zero and whose restriction to Ω is the largest non-negative harmonic function on Ω vanishing continuously on C and dominated by $M(|\Pi|\Omega|^q)$. The function u takes a positive value at each point of Ω. We thereupon introduce v with domain S whose restriction to $S - \Omega$ is the constant 1 and whose restriction to Ω is the smallest positive harmonic function on Ω having limit 1 at each point of C. If A is a sufficiently large positive number, then $u + Av$ is superharmonic on S and majorizes $|\Pi|^q$. We conclude that $\Pi \in H_q(S)$.

5. $O_p \prec \bar{O}_p$. The construction of this section is the most elaborate of the chapter. It combines the use of "Riemannian sectors" of the sort introduced with the aid of the E_{3+k} in §4, together with the attendant concern to cover a deleted neighborhood of ∞ by their Π images, and parameter control of the sort used in §3.

We take E_1 as in §4 distinguishing the same segments. We let (q_k) denote a decreasing sequence of positive numbers with limit p and (ε_k) an increasing sequence of positive numbers satisfying $\varepsilon_o = 1$ and $\lim \varepsilon_k = 2$. The choice of ε_o as 1 is not important. We let ν denote the least positive whole number exceeding $2q_o$. We take E_2 as a copy of $\Delta(0;2) - \{0\}$ for which the segments $[2^{-(2n+2)}, 2^{-(2n+1)}]$, $n = 0,1, \ldots,$ and the segments

$$\sigma_{jk} = [e^{2\Pi ij/\nu}\varepsilon_{2k}, e^{2\Pi ij/\nu}\varepsilon_{2k+1}],$$

$j = 0, \ldots, \nu - 1$, $k = 0,1, \ldots$ are distinguished. The points of E_2 will be, as above, of the form $(z,2)$.

The "Riemannian sectors". For $j = 0, \ldots, \nu - 1$, and $k = 0,1, \ldots,$ we introduce E_{jk} as the image with respect to

$$z \to (e^{2\Pi ij/\nu}\exp(\frac{\operatorname{Log} z}{q_k}), z, (j,k)) \tag{5.1}$$

of $\{\operatorname{Re} z > 0\}$ and distinguish the "parameter" segment

$$[(\varepsilon_{2k})^{q_k}, (\varepsilon_{2k+1})^{q_k}].$$

We define S, proceeding as in the earlier sections, joining E_2 to E_1 along the segments $[2^{-(2n+2)}, 2^{-(2n+1)}]$ and E_{jk} to E_2 along σ_{jk}, the precise technical sense of the procedure being that indicated in the earlier sections.

We also introduce for each whole number μ the approximate surface S_μ and its subregion S_μ^*. The surface S_μ is obtained by joining E_2 to E_1 as above and E_{jk} to E_2 but now only for (j,k) with $k = 0, \ldots, \mu$. The subregion S_μ^* is obtained by removing from S_μ the images with respect to $z \to (z,2)$ of the segments $\sigma_{jk}, k > \mu$. We see that (S_μ^*) is an increasing sequence of regions exhausting S.

We shall now show that independent of the choice of an allowed sequence (ε_k) the surface S is a member of $\bar{O}_p$ and that with a suitable choice of (ε_k) the projection map Π belongs to $H_p(S)$. We conclude that $O_p \prec \bar{O}_p$.

To show that $S \in \bar{O}_p$ we show that if $p < q < +\infty$, $S \in O_q$. Given such a q we fix an index l such that $q_l < q$. Now let $f \in H_q(S)$. A fortiori, $f \in H_{q_l}(S)$. A paraphrase of the argument of the preceding section shows that $f = g \circ \Pi$ where g is entire and by examining the behavior of f on the image of a sector $\{|\mathrm{Arg}\, z| < \alpha\Pi/2\} - \{\varepsilon_{2k}^{q_k}, \varepsilon_{2k+1}^{q_k}\}$, $0 < \alpha < 1$, with respect to (5.1), k being l, we conclude, as in §4, that $g(z) = O(|z|)$ and hence the constancy of f. Hence $S \in O_q$.

The parameter sequence (ε_k). We shall keep the ε_k with even indices and ε_1 fixed and shall choose the ε_{2k+3} recursively to obtain a sequence (S^*_μ) such that ($|\Pi|S^*_\mu|^p$ has a harmonic majorant and) the sequence $(M(|\Pi|S^*_\mu|^p))$ tends to a harmonic function on S in a non-decreasing fashion. It then follows that $\Pi \in H_p(S)$. We shall denote the projection map associated with S_μ by Π_μ.

We note that $|\Pi_\mu|^p$ has a harmonic majorant, $\mu = 0,1, \ldots$. To see this, it suffices to paraphrase the argument of the last paragraph of the preceding section. We let h_μ denote $M(|\Pi_\mu|^p)$ and show that <u>with</u> ε_k, $k = 0, \ldots, 2\mu+2$, <u>held fast</u> $h_{\mu+1}$ <u>tends pointwise to</u> h_μ <u>on</u> S_μ <u>less the points</u> $(e^{2\Pi ij/\nu}\varepsilon_{2\mu+2}, 2)$, $j = 0, \ldots, \nu - 1$, <u>as</u> $\varepsilon_{2\mu+3}$ <u>tends to</u> $\varepsilon_{2\mu+2}$. It is to be observed that the present situation involves a mild complication in comparison with that studied in §3 since the Π_μ are not bounded, however the special nature of the construction will assure success.

We begin by adapting aspects of the argument of §3 using some of the same notations Here C will be the union of the images with respect to the maps $z \to (e^{2\Pi ij/\nu}z, 2)$, $j = 0, \ldots, \nu - 1$ of $\{|z - \varepsilon_{2\mu+2}| = r\}$, r fixed, small and positive. We let Ω_1 denote the component of $S_{\mu+1} - C$ containing E_1 and we let Ω_2 denote $S_{\mu+1} - \bar{\Omega}_1$. In making the paraphrase we understand that $\varphi(x) = (x^+)^p$ and 'ν' is replaced by 'μ'. With this understanding we introduce U taken in the present context, save that here U has domain $S_{\mu+1}$. We let V denote the function on $S_{\mu+1}$ whose restriction to C is the constant 1 and whose restriction to Ω_k is the smallest harmonic function on Ω_k taking positive values and having limit 1 at each point of C, $k = 1,2$.

Before proceeding further it is desirable to obtain some information concerning the dependence of V on the parameter $\varepsilon_{2\mu+3}$. Let D denote the union of the images of

$$\Delta(\varepsilon_{2\mu+2};r) - \{\varepsilon_{2\mu+2}\}$$

with respect to the maps $z \to (e^{2\Pi ij/\nu}z,2)$ of the preceeding paragraph. It is easy to see that $V|D$ tends to 1 pointwise as $\varepsilon_{2\mu+3}$ tends to $\varepsilon_{2\mu+2}$.

We introduce A_j, the component of

$$\Pi_{\mu+1}^{-1}[\Delta(e^{2\Pi ij/\nu}\varepsilon_{2\mu+2};r)]$$

containing the point $(e^{2\Pi ij/\nu}\varepsilon_{2\mu+2},2)$, $j = 0, \ldots, \nu - 1$, and let A denote their union. We let Q denote the function with domain $S_{\mu+1}$ whose restriction to $\Omega_2 - \bar{A}$ is largest non-negative harmonic function on $\Omega_2 - \bar{A}$ majorized by

$$M(|\Pi_{\mu+1}|(\Omega_2 - \bar{A})|^p)$$

and having limit 0 at each point of $fr(\Omega_2 - \bar{A})$, and which takes the value 0 elsewhere on $S_{\mu+1}$. We let R denote the function with domain $S_{\mu+1}$ whose restriction to $fr(\Omega_2 - \bar{A})$ is the constant 1 and whose restriction to $S_{\mu+1} - fr(\Omega_2 - \bar{A})$ is the smallest positive harmonic function on this domain having limit 1 at each point of $fr(\Omega_2 - \bar{A})$. There exist positive numbers c and d, independent of $\varepsilon_{2\mu+3}$ near $\varepsilon_{2\mu+2}$ such that $Q + cR$ and $U + dV$ are superharmonic and satisfy

$$|\Pi_{\mu+1}|(\Omega_2 - \bar{A})|^p \leqslant (Q + cR)|(\Omega_2 - \bar{A})$$

and

$$|\Pi_{\mu+1}|\Omega_1|^p \leqslant (U + dV)|\Omega_1.$$

Hence we conclude that

$$Q + cR + U + dV$$

is a superharmonic majorant of $|\Pi_{\mu+1}|^p$ and therefore of $h_{\mu+1}$. As in §3, we let H with domain $S_\mu - \{(e^{2\Pi ij/\nu}\varepsilon_{2\mu+2},2) : j = 0, \ldots, \nu - 1\}$ be the pointwise limit of a sequence of $h_{\mu+1}$ (the parameter $\varepsilon_{2\mu+3}$ tending to $\varepsilon_{2\nu+2}$). Here we have

$$U(s) \leqslant h_\mu(s) \leqslant H(s) \leqslant \begin{cases} U(s) + dV(s), & s \in \bar{\Omega}_1; \\ d, s \in D. \end{cases}$$

It is to be observed that $R(s)$ tends to 0 as $\varepsilon_{2\mu+3}$ tends to $\varepsilon_{2\mu+2}$, $s \in S_\mu - \{(e^{2\Pi ij/\nu}\varepsilon_{2\mu+2}, 2) : j = 0, \ldots, \nu - 1\}$. The remainder of the argument now follows along the same lines as in §3. Thus we conclude the equality of $h_\mu(s)$ and $H(s)$ for admitted s and thereupon that $h_\mu(s)$ is the limit of $h_{\mu+1}(s)$ as $\varepsilon_{2\mu+3}$ tends to $\varepsilon_{2\mu+2}$, s admitted. We fix $a \in E_1$, repeat the last two paragraphs of §3, mutatis mutandis, and conclude that for S so defined, $\Pi \in H_p(S)$. Hence $S \notin O_p$.

The strict inclusion asserted at the beginning of this section follows.

6. Given $0 < p < q < +\infty$, we see on introducing s, $p < s < q$, that

$$\bar{O}_p \subset \underline{O}_s \prec \bar{O}_s \subset \underline{O}_q.$$

The inclusion relation (1.1) now follows on taking into account the results of §§2 - 4.

7. It would be of interest to determine whether the corresponding inclusions in the plane theory are strict. We have not succeeded in constructing appropriate examples.

The following remarks show that, at all events, there is a plane region in $O_1 - O_{LA}$. We let E be a compact subset of R having zero 1-dimensional Lebesgue measure but positive logarithmic capacity. We propose $\Omega = C - E$ as a candidate for membership in $O_1 - O_{LA}$. The identity map on Ω is Lindelöfian. (We note that Ω is hyperbolic.) Suppose that $f \in H_1(\Omega)$. Then the restrictions of f to the upper and lower half-planes belong to H_1 of the corresponding half-plane (in the sense of Ch.I.) and f has a removable singularity at ∞. Now $F_y : x \to f(x + iy), a \leqslant x \leqslant b$, converges in the mean of order 1 as $y \downarrow 0$ (resp. $y \uparrow 0$) and the respective limits in the mean are equal p.p. (in fact take the value $f(x)$ p.p.). Here $-\infty < a < b < +\infty$. Let σ be a segment of R containing E in its interior. We obtain with the aid of the Cauchy integral formula for $f(w), w \in C$ not a point of

$$\{\text{Re} z \in \sigma, |\text{Im } z| \leqslant h\},$$

h positive, on taking the limit as h tends to 0, a representation for $f(w), w \in C - \sigma$, from which we conclude that f is constant. It follows that $\Omega \in O_1 - O_{LA}$.

Chapter IV

Boundary Problems

1. In this chapter our principal concern will be the study of functions of various classes given on the border of a compact bordered Riemann surface. Thanks to the use of a Schottky doubling, we may and do assume that our compact bordered Riemann surface is the closure of a region Ω of a compact Riemann surface S such that Γ = frΩ is the union of a finite number (>o) of disjoint regular analytic closed Jordan curves and there exists a univalent anticonformal map α of S onto itself keeping each point of Γ fixed and mapping Ω onto $S-\bar{\Omega}$. Our program consists in obtaining representations for complex-valued functions in the class $L_p(\Gamma)$ in terms of functions of the class $H_p(\Omega)$ and related functions with domain $S-\bar{\Omega}$, and thereupon applying these representations to boundary questions involving Hardy classes pertaining to Ω. The representation to be developed is an extension of the classical one where $S=\hat{C}$, the extended plane, and $\Omega=\Delta$, which is a consequence of M.Riesz's conjugate series theorem and is standard in the classical study of Hardy classes [21]. The extension to be developed will be established with the aid of the classical representation to which we have just referred and of the so-called <u>unitary</u> meromorphic functions on S, i.e. meromorphic functions on S which are analytic at each point of $\bar{\Omega}$ and take values of modulus one at each point of Γ. The unitary functions reduce to just the finite Blaschke products in the classical case. It will be seen that the Schottky symmetry present in the setting will be very advantageous.

Parts of the material of this chapter were given in our paper [17].

2. The Theorem of M.Riesz (H_p version). Let $1 < p < +\infty$. Given $F \in L_p[0,2\Pi]$ and real-valued we let $\mu(F)$ denote

$$\left[\frac{1}{2\Pi} \int_0^{2\Pi} |F(\theta)|^p \, d\theta \right]^{j/p}.$$

The conjugate series theorem of M. Riesz states that for each such F the conjugate series of its Fourier series is the Fourier series of a function $\tilde{F} \in L_p[0,2\Pi]$ and that there exists a positive number C such that for all allowed F we have

$$\mu(\tilde{F}) \leqslant C\mu(F).$$

The infimum of such C is termed the <u>Riesz constant</u> associated with p. Its value for general p has not been determined.

The above theorem is equivalent to (and is easily derived from) the following theorem concerning $H_p(\Delta)$ which will be established by the argument of P.Stein [35]. Here $\| \ \|$ is taken in the sense of §2, Ch.I, with $S = \Delta$ and $q = 0$.

<u>Theorem 1</u> (M. Riesz H_p theorem): <u>Given</u> $1 < p < +\infty$. <u>Let</u> u <u>be a real-valued harmonic function on</u> Δ <u>such that</u> $|u|^p$ (<u>which is subharmonic on</u> Δ) <u>has a harmonic majorant. Let</u> f <u>be the unique analytic function on</u> Δ <u>satisfying</u> $\mathrm{Re} f = u$ <u>and</u> $f(0) = u(0)$. <u>Then for all such</u> u, <u>the function</u> f <u>is a member of</u> $H_p(\Delta)$. <u>Further there exists a positive number</u> C <u>such that</u>

$$\|f\| \leqslant C[(M|u|^p)(0)]^{1/p}$$

<u>for all allowed</u> u.

Before we turn to the proof we comment that the theorem of Szegö-Solomentsev assures us that u is the difference of quasi-bounded non-negative harmonic functions on Δ and that $M|u|^p$ is quasi-bounded. It is readily concluded that the allowed u are exactly the functions given by the Poisson-Lebesgue integrals

$$\frac{1}{2\Pi}\int_0^{2\Pi} U(\theta)k(\theta,z)d\theta, \quad |z| < 1.$$

where $U \in L_p[0,2\Pi]$ and is real-valued. These remarks show the close relation of the present question with the studies of Chapter II and indicate a connecting link between Theorem 1 and the M. Riesz conjugate series theorem.

Proof: The proof will be carried out in three stages. In the first we consider allowed u that are non-negative and p further restricted by the requirement $p \leq 2$. Here we consider only u taking strictly positive values, the excluded case being trivial. We let c denote a real number whose value will be restricted in the course of the argument. We introduce the auxiliary function

$$\theta = |f|^p - cu^p$$

(this is one of the key steps of the argument of P. Stein) and obtain by elementary calculation

$$\Delta\theta = 4\theta_{z\bar{z}} = p^2|f|^p\left|\frac{f'}{f}\right|^2 - p(p-1)cu^{p-2}|f'|^2$$

$$= p^2|f'|^2[|f|^{p-2} - (1 - \frac{1}{p})cu^{p-2}]$$

$$\leq p^2|f'|^2u^{p-2}[1 - (1 - \frac{1}{p})c].$$

We take $c = p/(p-1)$ and observe that θ is superharmonic on Δ. Consequently $|f|^p$ has as a superharmonic majorant

$$\theta + \frac{p}{p-1} M(u^p)$$

and

$$\|f\|^p \leq \theta(0) + \frac{p}{p-1} M(u^p)(0)$$

$$< \frac{p}{p-1} M(u^p)(0).$$

The theorem is established for allowed non-negative u when $1 < p \leq 2$.

We now show that the requirement that u be non-negative may be dropped, the condition $1 < p \leq 2$ persisting. To that end we introduce $U_1 = M(u^+)$, $U_2 = M(-u^-)$, F_k the analytic function on Δ with real part U_k satisfying $F_k(0) = U_k(0)$, $k = 1,2$. Noting that

$$|u|^p = (u^+)^p + (-u^-)^p,$$

we conclude with the aid of Th. 2, Ch. II that U_1^p and U_2^p have harmonic majorants and that

$$M(|u|^p) = M(U_1^p) + M(U_2^p).$$

Using the result of the preceeding paragraph we see that $f = F_1 - F_2 \in H_p(\Delta)$. We now conclude that

$$\begin{aligned} \|f\|^p &\leq (\|F_1\| + \|F_2\|)^p \\ &\leq 2^{p-1}(\|F_1\|^p + \|F_2\|^p) \\ &\leq 2^{p-1} \frac{p}{p-1} M(|u|^p)(0). \end{aligned}$$

The assertion of the first sentence of this paragraph is established.

There remains the final stage where $2 < p < +\infty$ and u is not restricted with respect to sign. It will be referred back to the case where p is replaced by $p/(p-1)$ with the aid of the Cauchy integral formula and well-known properties of linear functionals on L_p spaces. We let v denote the imaginary part of f. We consider a polynomial function whose imaginary part V takes the value 0 at 0. We denote its real part by U. Let $0 < r < 1$. By the Cauchy integral formula we obtain

$$\int_0^{2\Pi} u(re^{i\theta})V(e^{i\theta})\,d\theta = -\int_0^{2\Pi} v(re^{i\theta})U(e^{i\theta})\,d\theta$$

and thereupon with the aid of the Hölder inequality and the results of the preceding paragraph we conclude that

$$\begin{aligned} &\left|\frac{1}{2\Pi}\int_0^{2\Pi} v(re^{i\theta})U(e^{i\theta})\,d\theta\right| \\ &\leq \left[\frac{1}{2\Pi}\int_0^{2\Pi} |u(re^{i\theta})|^p d\theta\right]^{1/p} \left[\frac{1}{2\Pi}\int_0^{2\Pi} |V(e^{i\theta})|^{\frac{p}{p-1}} d\theta\right]^{\frac{p-1}{p}} \end{aligned}$$

$$\leqslant [M(|u|^p)(0)]^{1/p} \, C[M(|U|^{\frac{p}{p-1}})(0)]^{\frac{p-1}{p}} ,$$

where C is independent of allowed u and U. We now conclude using well-known facts concerning the norm of linear functionals on L spaces and the possibility of approximating members of $L_{p/(p-1)}[0,2\Pi]$ in the mean of order $p/(p-1)$ by allowed $U(e^{i\theta})$ that

$$\left[\frac{1}{2\Pi}\int_0^{2\Pi} |v(re^{i\theta})|^p d\theta\right]^{1/p}$$

$$\leqslant C[M(|u|^p)(0)]^{1/p} .$$

Thereupon we conclude with the aid of the Minkowski inequality that

$$\frac{1}{2\Pi}\int_0^{2\Pi} |f(re^{i\theta})|^p d\theta \leqslant (1 + C)^p M(|u|^p)(0).$$

It follows that $f \in H_p(\Delta)$ and that

$$\|f\|^p \leqslant (1 + C)^p M(|u|^p)(0).$$

Theorem 1 is now established.

It is to be remarked that the exclusion of the cases $p = 1$ and $p = +\infty$ was not accidental. The theorem does not hold for these cases.

Our object in introducing Theorem 1 is to prepare the way for the representation of a complex-valued member of $L_p[C(0;1)]$, $1 < p < +\infty$, as the sum p.p. of Fatou boundary functions belonging to $H_p(\Delta)$ and $H_p[\Delta(\infty;1)]$. [$C(a;r)$ denotes the circumference with center a, radius r. $\Delta(\infty;r)$ is the set of z in the extended plane satisfying $|z| > r^{-1}$.] We begin with a uniqueness theorem which is valid in the H_1 setting.

<u>Theorem</u> 2: <u>Let</u> $f \in H_1(\Delta)$, $g \in H_1[\Delta(\infty;1)]$, $g(\infty) = 0$. <u>If</u> $f^*(z) + g^*(z) = 0$p.p. <u>on</u> $C(0;1)$, <u>then</u> f <u>and</u> g <u>are identically zero</u>.

This result may be arrived at almost immediately on noting that

$$\int_0^{2\Pi} f^*(e^{i\theta})e^{ki\theta}d\theta = 0, \ k = 1,2, \ldots,$$

and

$$\int_0^{2\Pi} g^*(e^{i\theta})e^{ki\theta}d\theta = 0, \ k = 0,-1,-2, \ldots.$$

so that all the Fourier coefficients of $f^*(e^{i\theta})$ and $g^*(e^{i\theta})$ are 0 as a consequence of the hypothesis of the theorem. However it is desirable to have a proof that extends conveniently to the Riemann surface situation, which is our principal concern in this chapter. To that end we proceed by showing that there exists a function analytic on the extended plane whose restriction to Δ is f and whose restriction to $\Delta(\infty;1)$ is $-g$, an assertion from which the theorem follows at once. We shall use the Cauchy integral theorem and Cauchy formula for an annulus together with the mean convergence of $f(re^{i\theta})$ to $f^*(e^{i\theta})$ as $r\uparrow 1$ and the corresponding property of g in order to exhibit an exact prototype argument for the Riemann surface situation. We recall the convergence in the mean of order p of $f(re^{i\theta})$ to $f^*(e^{i\theta})$ for $f \in H_p(\Delta)$, $1 \leq p < +\infty$; it may be demonstrated simply with the aid of the Poisson-Lebesgue representation of f. [F. Riesz [31; p.651] showed that this result is valid for all positive p.] Let $0 < r < 1$. [We shall consider r near 1 in the Riemann surface situation.] We have

$$f(z) = \frac{1}{2\Pi i}\oint_{C(0;1)} \frac{f^*(\zeta)}{\zeta - z}\,d\zeta - \frac{1}{2\Pi i}\oint_{C(0;r)} \frac{f(\zeta)}{\zeta - z}\,d\zeta$$

$$= -\frac{1}{2\Pi i}\oint_{C(0;1)} \frac{g^*(\zeta)}{\zeta - z}\,d\zeta - \frac{1}{2\Pi i}\oint_{C(0;r)} \frac{f(\zeta)}{\zeta - z}\,d\zeta$$

$$= -\frac{1}{2\Pi i}\oint_{C(0;r^{-1})} \frac{g(\zeta)}{\zeta - z}\,d\zeta - \frac{1}{2\Pi i}\oint_{C(0;r)} \frac{f(\zeta)}{\zeta - z}\,d\zeta ,$$

$r < |z| < 1$. An analogous argument shows that the value of the last line of the display is $-g(z)$, $1 < |z| < r^{-1}$. We conclude that f and $-g$ are the restrictions to their respective domains of a function analytic on the extended plane.

It is now clear that if f_k (resp. g_k) satisfies the condition imposed on f

(resp. g), $k = 1,2$, and if in addition

$$f_1^*(z) + g_1^*(z) = f_2^*(z) + g_2^*(z)$$

p.p. on $C(0;1)$, then $f_1 = f_2$ and $g_1 = g_2$.

We now turn to the existence aspect of the question and here restrict attention to the case where $1 < p < +\infty$ as is inevitable. The basic result is the following

Theorem 3: Let $1 < p < +\infty$. (a) If $F \in L_p[C(0;1)]$, then there exists $f \in H_p(\Delta)$, $g \in H_p[\Delta(\infty;1)]$, $g(\infty) = 0$, such that

$$F(z) = f^*(z) + g^*(z)$$

p.p. on $C(0;1)$. The ordered pair (f,g) is uniquely determined by these requirements. (b) There exists a positive number C such that for each allowed F we have

$$\|f\|, \|g\| \leq C \left[\frac{1}{2\pi}\int_0^{2\pi} |F(e^{i\theta})|^p d\theta\right]^{1/p}.$$

Here $\|g\|$ is taken relative to $S = \Delta(\infty;1), q = \infty$.

The uniqueness aspect of the theorem is already cared for. The proof of the remaining assertions is an easy consequence of Theorem 1, this §. We first show that the case of complex-valued F may be reduced to that of real-valued F. We let F_1 denote the real part of F and F_2 the imaginary part. We let (f_k,g_k) denote the ordered pair associated in the theorem with F_k $(k = 1,2)$. Then the ordered pair (f,g) given by

$$f = f_1 + if_2,\ g = g_1 + ig_2,$$

serves for F. It is immediate that f and g fullfill the requirements imposed in (a). If C serves in (b) for the real case, we see that

$$\|f\| \leq \|f_1\| + \|f_2\|$$

$$\leq C\left[\frac{1}{2\pi}\int_0^{2\pi}|F_1(e^{i\theta})|^p d\theta\right]^{1/p} + C\left[\frac{1}{2\pi}\int_0^{2\pi}|F_2(e^{i\theta})|^p d\theta\right]^{1/p}$$

$$\leq 2C\left[\frac{1}{2\pi}\int_0^{2\pi}|F(e^{i\theta})|^p d\theta\right]^{1/p}$$

and that the corresponding inequalities hold for g.

To treat the case of real-valued F we associate with F the Poisson-Lebesgue integral

$$u(z) = \frac{1}{2\pi}\int_0^{2\pi} F(e^{i\theta})k(\theta,z)d\theta, \quad |z| < 1,$$

and let w denote the analytic function with domain Δ satisfying $\mathrm{Re}\, w = u$, $w(0) = u(0)$. For the functions f and g one proposes the functions given by

$$f(z) = [w(z) + \overline{w(0)}]/2,$$

$$g(\bar{z}^{-1}) = [\overline{w(z)} - \overline{w(0)}]/2,$$

$|z| < 1$. Since $F \in L_p[C(0;1)]$, we are assured that $|u|^p$ has a harmonic majorant (Hölder inequality). It is now routine to check with the aid of Theorem 1 that Theorem 3 is valid for the case of real-valued F.

3. Minimal positive harmonic functions. We return to the setting of §2,Ch.I. Our object is to introduce apparatus that will furnish existence results concerning unitary functions wanted in the following section as well as information concerning minimal positive harmonic functions which will be of use on several occasions.

We recall that a positive harmonic function u on S is termed minimal provided that whenever v is a positive harmonic function on S satisfying $v \leq u$, then v/u is constant. The notion was introduced by R.S. Martin [23] in 1941 and has turned out to be of fundamental importance in the theory of harmonic functions. cf. [5]. It is readily verified that $u \in Q$ is minimal if and only if u is an extreme point of Q.

The following lemma serves as a starting point of the developments of this section.

Lemma 1: Let λ denote a continuous positive homogeneous, additive map of P, the set of positive harmonic functions on S, into R^n, n a positive whole number. Then each point of $\lambda(Q)$ is the image with respect to λ of the barycenter of at most $n + 1$ extreme points of Q.

Gloss: To say that λ is positive homogeneous means that given c a positive number and $u \in P$, $\lambda(cu) = c\lambda(u)$; to say that λ is additive means that given $u, v \in P$, $\lambda(u+v) = \lambda(u) + \lambda(v)$. We understand that the space of continuous real-valued functions on S is endowed with the topology generated by the sets $N(f,K,\varepsilon)$ consisting of the continuous real-valued functions g on S satisfying

$$\max_{s \in K} |g(s) - f(s)| < \varepsilon,$$

where f is a continuous real-valued function on S, K is a compact subset of S, and ε is a positive real number. The continuity of λ is referred to the relative topology on P.

On noting that Q is a convex and compact subset of the space of continuous real-valued functions on S, that by the elementary theory of convex sets in finite dimensional euclidean spaces each point of $\lambda(Q)$ is the barycenter of at most $n + 1$ extreme points of $\lambda(Q)$ [cf. 4.p.15], and that the set of $u \in Q$ mapped by λ into an extreme point of $\lambda(Q)$ is a closed support of Q [cf. 22. p. 130], we conclude Lemma 1 with the aid of the Krein-Milman theorem [cf. 22. p. 131].

This lemma was given in [14]. In that paper a proof was given without appeal to the Krein-Milman theorem. We indicate the argument. It is readily seen that the proof of Lemma 1 reduces to showing that each extreme point of $\lambda(Q)$ is the image of an extreme point of Q. This may be achieved as follows. Let (s_k) be a univalent sequence of points in S whose image is dense in S. Let e be a given extreme point of $\lambda(Q)$. There exists a sequence of functions in Q, say (u_k), satisfying the following conditions: $\lambda(u_o) = e$; for each whole number k, u_{k+1} is a member of the set E_k of $u \in Q$ satisfying $\lambda(u) = e$, $u(s_j) = u_k(s_j)$, $j = 0, \ldots, k-1$, which in addition satisfies

$$u_{k+1}(s_k) = \max_{u \in E_k} u(s_k).$$

Clearly, (u_k) possesses a limit, say v, and $\lambda(v) = e$. Further v is an extreme point of Q. To see this we proceed as follows. If $v = (1-t)v_1 + tv_2$ where $0 < t < 1$ and $v_1, v_2 \in Q$, then $\lambda(v_1) = \lambda(v_2) = e$. If $v_1 \neq v$, there is a least whole number k such that $v(s_k) \neq v_1(s_k)$, say 1. But then $v(s_1) < \max\{v_1(s_1), v_2(s_1)\}$ and $v_1, v_2 \in E_1$. The condition: $u_{1+1}(s_1) = \max_{u \in E_1} u(s_1)$ is violated. Hence $v_1 = v = v_2$.

An immediate consequence of the existence of such a v associated with an extreme point of $\lambda(Q)$ is the existence of a minimal positive harmonic function on S as we see on considering the special case $\lambda(u) = u(a)$.

Henceforth in this section we shall be concerned with regions Ω defined in §1, this Ch., and we examine first of all the question of the boundary behavior of minimal positive harmonic functions on Ω. The situation is like the classical one for the unit disk where the minimal positive harmonic functions are just the functions $z \longrightarrow ck(\theta, z)$, c a positive number.

<u>Theorem</u> 4: <u>Let</u> u <u>be a positive harmonic function on</u> Ω. <u>Then</u> u <u>is minimal if and only if</u> u <u>has limit</u> 0 <u>at all but one point of</u> $\Gamma = \text{fr}\Omega$. <u>For</u> u <u>minimal, if</u> b <u>is the point of</u> Γ <u>at which</u> u <u>does not have limit</u> 0, <u>then the harmonic prolongation of</u> u <u>to</u> $S - \{b\}$ <u>is given locally by</u> $\text{Re}(z^{-1})$ <u>in terms of a suitable uniformizer of</u> S <u>taking</u> 0 <u>into</u> b.

Proof: Suppose that u is positive harmonic on Ω and has limit 0 at each point of Γ save b. We show that if v is also a positive harmonic function on Ω having limit 0 at each point of Γ save b, then v is proportional to u. The minimality of u then follows. To that end we make use of the well-known fact that a function h

harmonic on $\{|z| < 1, \operatorname{Im} z > 0\}$ which takes positive values and satisfies $\lim_x h = 0$ for x real, $0 < |x| < 1$, admits a representation of the form

$$h(z) = \operatorname{Re}(i\alpha z^{-1}) + h_1(z)$$

where α is a non-negative real number and h_1 is harmonic on Δ and takes the value 0 at each point of the interval $]-1,1[$. Taking a uniformizer θ for S with domain Δ which takes 0 into b, the open upper semicircle into a subset of Ω, and the complement into a subset of $S - \bar{\Omega}$, we see on composing u and v with the restriction of θ to $\{|z| < 1, \operatorname{Im} z > 0\}$ that there exists a positive number c such that $v - cu$ has the limit 0 at each point of Γ. Hence $v = cu$.

Suppose now that u is a minimal positive harmonic function on Ω. Let b_1 and b_2 be distinct points of Γ. Let θ_k be uniformizers of the type described in the preceeding paragraph taking 0 into b_k, $k = 1,2$, and having disjoint images. We let u_k denote the function with domain Ω whose restriction to $\delta_k = \theta_k(\{|z| < 1/2, \operatorname{Im} z > 0\})$ is the largest non-negative harmonic function with domain δ_k dominated by the restriction of u to δ_k and having limit 0 at each point of $\theta_k(\{|z| = 1/2, \operatorname{Im} z > 0\})$ and which takes the value 0 at each point of $\Omega - \delta_k$, $k = 1,2$. It is readily verified that $u - u_k$ has limit 0 at $\theta_k(x)$, $-1/2 < x < 1/2$. In fact, if v_k is the smallest positive harmonic function on δ_k having limit $u[\theta_k(e^{i\varphi}/2)]$ at $\theta_k(e^{i\varphi}/2)$, $-\Pi/2 < \varphi < \Pi/2$, then v_k has limit 0 at $\theta_k(x)$, $-1/2 < x < 1/2$ and $u_k(s) = u(s) - v_k(s)$, $s \in \delta_k$, $k = 1,2$. The functions u_k are subharmonic. Now Mu_1 is proportional to u by the minimality of u. Also Mu_1 has limit 0 at b_2 as we see on noting that $u_1 \leqslant u - u_2$. Suppose that u does not have limit 0 at either b_1 or b_2. Then Mu_1 is positive since u_1 is not the constant 0. Hence u is a constant positive multiple of Mu_1 and so has limit 0 at b_2. Contradiction. We conclude that u has limit 0 at all but one point of Γ. Of course, there is a point of Γ at which u does not have limit 0. The last assertion of the theorem is immediate.

Unitary functions. We recall that the notion of a unitary function was defined in §1, this Ch..] We fix $a \in \Omega$ as the normalization point for Q in the present setting. We let T denote the Möbius transformation satisfying

$$T(z) = \frac{z-1}{z+1}, \quad z \neq \infty.$$

Let the one-dimensional homology group of Ω be trivial. We are assured of the existence of a minimal positive harmonic function u on Ω by earlier results of this section. By the triviality of the one-dimensional homology group and the information of Theorem 4 we see that u is the real part of the restriction to Ω of a function f meromorphic on S (of §1, this Ch.), having a simple pole at some point b of Γ, analytic elsewhere, and such that $\operatorname{Re} f(s) = 0$, $s \in \Gamma - \{b\}$. It is immediate that $T \circ f$ is unitary and not constant.

Suppose now that the one-dimensional homology group of Ω is not trivial and that $\gamma_1, \ldots, \gamma_m$ is a basis. We let $\omega_k(u)$ denote the period associated with γ_k of the abelian differential δu given locally in terms of uniformizers θ by $2(u \circ \theta)_z dz$, u being harmonic on Ω. The map

$$u \to \lambda(u) = (\omega_1(u), \ldots, \omega_m(u))$$

is admissible in the sense of Lemma 1. We conclude that since the constant 1 belongs to Q, there is a member u of Q which is the sum of at most $m + 1$ minimal positive harmonic functions on Ω and satisfies $\omega_k(u) = 0$, $k = 1, \ldots, m$. The final part of the argument of the preceding paragraph goes over with obvious modifications, the vanishing of the $\omega_k(u)$ assuring that u is the real part of an analytic function on Ω.

Interpolation questions. Another application of the arguments just employed is the following:

Let f be an analytic function on Ω having modulus less than one. Let $s_1, \ldots, s_n$ be $n (\geqslant 1)$ distinct points of Ω. Let θ_k be a uniformizer satisfying $\theta_k(0) = s_k$ and let ν_k be a whole number, $k = 1, \ldots, n$. The question arises whether there exists a unitary function g satisfying

$$(g \circ \theta_k)^{(j)}(0) = (f \circ \theta_k)^{(j)}(0), \tag{3.1}$$

$j = 0,\ldots,\nu_k$; $k = 1,\ldots,n$. The answer, which is affirmative, as we shall see, furnishes a useful approximation tool.

We first note that it suffices to treat the corresponding problem of associating with analytic functions having positive real part analytic functions having real part the sum of a finite number of minimal positive harmonic functions and satisfying interpolating conditions corresponding to (3.1). Indeed, we take a distinct from the s_k and now let T denote a Möbius transformation mapping Δ onto the right-half plane and satisfying $T[f(a)] = 1$. We let $F = T \circ f$ and seek to show that there exists G analytic on Ω, $\mathrm{Re}\, G$ the sum of a finite number of minimal positive harmonic functions on Ω such that (3.1) holds with F replacing f and G replacing g, for then $T^{-1} \circ G$ is the restriction to Ω of a unitary function g satisfying (3.1). To that end we introduce a map λ where $\lambda(u)$ has $m + 2n + 2\Sigma\nu_k$ components consisting of the following: the $\omega_k(u)$, $k = 1,\ldots,m$; $u(s_k)$, $k = 1,\ldots,n$; the imaginary parts of the integrals of δu along fixed paths in Ω joining a to s_k, $k = 1,\ldots,n$; the real and imaginary parts of $2(u \circ \theta_k)_z^{(j)}(0)$, $j = 0,\ldots,\nu_k - 1$, $k = 1,\ldots,n$. On taking u as the sum of a finite number of minimal positive harmonic functions such that $u(a) = 1$ and

$$\lambda(u) = \lambda(\mathrm{Re}\, F),$$

we see that u is the real part of an analytic function G satisfying $G(a) = 1$. Thanks to this normalization, it is seen that G satisfies (3.1) with F replacing f and G replacing g.

It is to be remarked that this result yields qualitative information concerning the Pick-Nevanlinna interpolation problem for bounded analytic functions in the setting of Riemann surfaces having finite topological characteristics and non-degenerate (i.e. not pointlike boundary components: if interpolation conditions, finite in number, are fulfilled by some analytic function on the surface of modulus less than one, then the same conditions are fulfilled by an analytic function of modulus less than one whose modulus has limit 1 at the adjoined point of the Alexandroff compactification of the given Riemann surface.

4. Provisional decomposition theorem for $L_p(\Gamma), 1 < p < +\infty$. Our principal concern is the extension of Theorem 3 of this chapter to the L_p class associated with Γ appropriately defined. In this section we obtain a first provisional theorem in this direction (Theorem 5) which will serve to help derive the final version (Theorem 8) to be treated in §6 of this chapter. We emphasize that the remainder of this chapter continues in the setting put down in §1.

$L_p(\Gamma)$. We fix $a \in \Omega$ and let $\mathcal{g}_a$ denote Green's function for Ω with pole a as well as its prolongation by Schwarzian reflexion to S. We introduce the measure

$$\frac{1}{2\pi}\frac{\partial \mathcal{g}_a}{\partial n}\,ds$$

on Γ where the derivative is construed as the inner normal derivative relative to Ω. $L_p(\Gamma)$ is taken as the class of complex-valued functions which are measurable and such that the pth power of the modulus is integrable. Of course, the notion can be defined without reference to a and the induced measure by the use of uniformizers, but the question is not important for our purposes. We fix $p, 1 < p < +\infty$, and define $N(f)$ for $f \in L_p(\Gamma)$ by

$$N(f) = \left(\frac{1}{2\pi}\int_\Gamma |f|^p \frac{\partial \mathcal{g}_a}{\partial n}\,ds\right)^{1/p}. \tag{4.1}$$

The genus of S will be denoted by g.

Given f meromorphic on a Riemann surface T by the divisor of f, ∂_f, we understand the constant $+\infty$ on T when f is the constant 0 and otherwise the function with domain T assigning to each point $s \in T$ the common value of the minimum of the indices k for which the kth O-Laurent coefficient of $f \circ \theta$ is not 0, θ being a uniformizer, $\theta(0) = s$. [To be precise, by the kth O-Laurent coefficient of g analytic in some deleted neighborhood of 0 we understand the common value of the kth Laurent coefficient of $g|[\Delta(0;r) - \{0\}]$, r small and positive.] Given an abelian differential ω on T, the divisor ∂_ω is correspondingly defined. Given T compact, a point b is said to be a <u>Weierstrass point</u> of T provided that there exists a non-constant meromorphic function f, analytic at each point of $T - \{b\}$ and such that $\partial_f(b)$ is at least as large as the negative of the genus of T. The Weierstrass points are finite in number. They are absent for surfaces of genus 0 or 1 and are present for surfaces of genus greater than 1. If b is not a Weierstrass point, then for each whole number n exceeding the genus of T, there exists a meromorphic function on T with pole of order n at b which has no poles elsewhere. If b is a Weierstrass point, the conclusion of the preceding sentence holds when n is at least as large as twice the genus of T. An excellent account of the theory of Weierstrass points is given in the treatise of Behnke-Sommer[2].

Discriminant. Let φ and ψ be meromorphic on T, which we continue to suppose compact in this and the following paragraph. We suppose that φ is not constant and we let n denote the common value of its valence. By <u>the discriminant of ψ with respect to φ</u>, denoted $D_\varphi[\psi]$, is meant the rational function taking as its value at each point z of the extended plane which has n distinct pre-images with respect to φ, say $s_1,\dots,s_n$, and is such that $\psi(s_k) \neq \infty$, $k = 1,\dots,n$, the square of

$$\det\{[\psi(s_k)]^{j-1}\}.$$

The following fact will be useful: Given b,c,d distinct Points of T, there exists a meromorphic function on T having a pole at b but nowhere else, and taking distinct values at c and d. Indeed, there exists a function f analytic on $T - \{b\}$ which has a zero at c and no others. This fact may be shown with the aid of the Weierstrass gap theorem [2] and the use of appropriate generating harmonic functions or alternatively with the aid of the theorem of Behnke and Stein [3], which constitutes an appeal to deeper results. There are meromorphic functions φ and ψ on T having poles at b of respective orders n, $n + 1$, where n exceeds twice the genus of T, which are elsewhere analytic. It is classical that $D_\varphi[\psi]$ is not the constant zero. It follows from standard considerations that

$$f = \sum_{0}^{n-1} (A_k \circ \varphi_o)\psi_o^{\ k},$$

where the A_k are meromorphic on C and have a finite number of poles, and φ_o (resp. ψ_o) denotes the restriction to $T - \{b\}$ of φ (resp. ψ). Indeed the fact that $D_\varphi[\psi]$ is not the constant zero permits us to establish this assertion with the aid of the elementary theory of systems of linear equations. On approximating the A_k by rational functions B_k with the same principal parts as A_k at each point of C, we see that such B_k may be so chosen that

$$\sum_{0}^{n-1} (B_k \circ \varphi)\psi^k,$$

which has a pole at most at b, takes distinct values at c and d and so is a function of the desired type.

We return to S and fix a non-constant unitary function u on S. We fix $b \in S - \bar{\Omega}$ and show

<u>Lemma</u> 2: <u>There exists</u> w <u>meromorphic on</u> S <u>having a pole at</u> b <u>but nowhere else</u> <u>such that</u> $D_u[w]$ <u>does not take the value</u> 0 <u>at any point of</u> $C(0;1)$.

Proof: We put aside the trivial case where the common value n of the valence of u is 1. At all events, the set of w meromorphic on S having a pole at b but nowhere else and such that $D_u[w]$ is not the constant 0 is not empty. To see this let $z \in \hat{C} - \{u(b)\}$ be chosen so that it has n distinct preimages $s_1,\dots,s_n$ with respect to u. There exists a meromorphic function on S having a pole at b but nowhere else which takes distinct values at the s_k. Indeed, there is a meromorphic function on S having a pole at b but nowhere else taking respectively the values 1 and 0 at two given distinct preimages of z with respect to u, thanks to the developments before the introduction of Lemma 2. By taking products of such functions we obtain a function taking the value 1 at a given preimage of z and the value 0 at the others which too is meromorphic on S and has a pole exactly at b. Taking a linear combination of functions of the type just obtained with distinct coefficients, one such function for each preimage of z we achieve the construction of a meromorphic function on S having a pole exactly at b and taking distinct values at the s_k. The discriminant of the so constructed function with respect to u takes a non-zero value at z and consequently is not the constant zero.

For each w meromorphic on S having a pole at b but nowhere else and such that $D_u[w]$ is not the constant zero we introduce $\nu(w)$, the sum of the multiplicities of the zeros of $D_u[w]$ on $C(0;1)$. Let w_o be an allowed w minimizing $\nu(w)$. Lemma 2 will be established by showing that $\nu(w_o) = 0$.

Suppose that $\nu(w_o) > 0$. We fix r, $0 < r < 1$, so that $D_u[w_o]$ is analytic at each point of $\{r \leqslant |z| \leqslant r^{-1}\}$ and the sum of the multiplicities of the zeros of $D_u[w_o]$ belonging to $\{r \leqslant |z| \leqslant r^{-1}\}$ is $\nu(w_o)$. Let $\zeta \in C[0;1]$ be a zero of $D_u[w_o]$. There exist distinct $c, d \in u^{-1}(\{\zeta\})$ such that $w_o(c) = w_o(d)$. We introduce φ meromorphic on S, having a pole at b but nowhere else and such that $\varphi(c) \neq \varphi(d)$. We show that a non-zero complex number t may be so chosen that

$$\nu(w_o + t\varphi) < \nu(w_o),$$

whence we obtain a contradiction. At the least we know that when t is small, the sum of the multiplicities of the zeros of $D_u[w_o + t\varphi]$ in $\{r \leqslant |z| \leqslant r^{-1}\}$ is $\nu(w_o)$. Our object will be achieved when we show that a small t may be so chosen that $D_u[w_o + t\varphi]$ has a zero in $\{r \leqslant |z| \leqslant r^{-1}\}$ not of modulus 1. We take a positive number ϱ so small that

$$\Delta(\zeta;\varrho) \subset \{r < |z| < r^{-1}\}$$

and that there exist local analytic inverses of u, namely σ and τ, with domain

$\Delta(\zeta;\varrho)$ which satisfy $\sigma(\zeta) = c$, $\tau(\zeta) = d$, and are such that $\varphi \circ \tau - \varphi \circ \sigma$ has no zero. It is to be observed that u has multiplicity 1 at each point of Γ. The function

$$\omega = \frac{w_o \circ \sigma - w_o \circ \tau}{\varphi \circ \tau - \varphi \circ \sigma}$$

has a zero at ζ, but it is not the constant zero. There exist arbitrarily small t in

$$\omega[\Delta(\zeta;\varrho)] - \omega[\Delta(\zeta;\varrho) \cap C(0;1)]. \tag{4.2}$$

Let t be a point of (4.2) and suppose that $t = \omega(z)$. Then $|z| \neq 1$. Since

$$w_o[\sigma(z)] + t\varphi[\sigma(z)] = w_o[\tau(z)] + t\varphi[\tau(z)],$$

it follows that

$$D_u[w_o + t\varphi](z) = 0.$$

The proof of the lemma follows on taking t in the set (4.2) sufficiently small.

Boundary behavior of members of $H_p(\Omega)$. We start with some preliminaries and note that if $F \in H_p(\{r_1 < |z| < r_2\})$, $0 < r_1 < r_2 < +\infty$, then $F_1 \in H_p[\Delta(0;r_2)]$ and $F_2 \in H_p[\Delta(\infty;r_1^{-1})]$, where F_1 and F_2 are the components of the Laurent decomposition of F : F_1 analytic on $\Delta(0;r_2)$, F_2 analytic on $\Delta(\infty;r_1^{-1})$, $F_2(\infty) = 0$, $F(z) = F_1(z) + F_2(z)$, $r_1 < |z| < r_2$. This assertion is in fact valid when $0 < p \leq +\infty$. The case $p = +\infty$ is immediate and will not enter in the discussion which follows. We introduce $r, r_1 < r < r_2$, and h a harmonic majorant of $|F|^p$. We use a representation of the form $h(z) = h_1(z) + h_2(z)$, where h_1 is harmonic on $\Delta(0;r_2)$ and h_2 is harmonic on $\{r_1 < |z| < +\infty\}$, to conclude that

$$\begin{aligned} |F_1(z)|^p &\leq 2^p[|F(z)|^p + |F_2(z)|^p] \\ &\leq 2^p h_1(z) + A, \end{aligned}$$

$r \leq |z| < r_2$, where A is a suitably chosen positive number. It follows by the subharmonicity of $|F_1|^p$ that $|F_1|^p \leq 2^p h_1 + A$. Hence $F_1 \in H_p[\Delta(0;r_2)]$. We conclude that F possesses Fatou boundary values p.p. on $C(0;r_2)$. We may even conclude for $0 < p < +\infty$ that $F(re^{i\theta})$ tends in the mean of order p to $F^*(r_2 e^{i\theta})$ as r tends to r_2 thanks to the fact that the corresponding statement holds for F_1 by the result of F. Riesz cited earlier [31, p. 651].

On introducing a univalent conformal map θ of an annulus $\{\varrho < |z| < \varrho^{-1}\}$ into S mapping $C(0;1)$ onto a component of Γ and mapping points in its domain of modulus less than 1 into points of Ω, we conclude from the fact that a member of $H_p(\Omega)$ composed with θ restricted to $\{\varrho < |z| < 1\}$ belongs to $H_p(\{\varrho < |z| < 1\})$ that a member of $H_p(\Omega)$ has a Fatou boundary function $\in L_p(\Gamma)$, $0 < p < +\infty$. The same conclusion holds when Ω is replaced by $\alpha(\Omega)$.

We now fix b to be a non-Weierstrass point of S lying in $\alpha(\Omega)$. The stipulation that b be a non-Weierstrass point is a matter of technical convenience. We fix w satisfying the requirements stated in Lemma 2 of this section. We introduce two vector spaces of meromorphic functions on S. The first, $\mathfrak{G}$, consists of the meromorphic functions on S whose divisors take values not less than $-g$ at b and $\alpha(b)$ and elsewhere not less than 0. The dimension of $\mathfrak{G}$ as a vector space over C is $g + 1$. The second, $\mathfrak{b}$, consists of the meromorphic functions on S having a pole at most at b and taking the value 0 at a. The space $\mathfrak{b}$ will enter the proof but not the statement of Theorem 5. The notation "N" below is given by (4.1), this Ch..

<u>Theorem</u> 5: <u>Let</u> $1 < p < +\infty$. (a) <u>If</u> $f \in L_p(\Gamma)$, <u>then</u>

$$f(s) = f_1^*(s) + f_2^*(s) + \sigma(s)$$

<u>p.p. on</u> Γ <u>for exactly one</u> (f_1, f_2, σ) <u>where</u> $f_1 \in H_p(\Omega)$, $f_2 \in H_p[\alpha(\Omega)]$, $\sigma \in \mathfrak{b}$, $f_1(a)=0$ $f_2[\alpha(a)] = 0$. <u>Here * refers to the Fatou boundary function</u>. (b) <u>There exists a positive number</u> C <u>such that for each</u> $f \in L_p(\Gamma)$, <u>we have</u>

$$N(f_1^*), N(f_2^*), N(\sigma|\Gamma) \leq CN(f).$$

The existence part of the following proof uses the suggestion, to start with (4.3) for $f \in L_p(\Gamma)$ and to make the Riesz decomposition of the A_k at the outset, made to me by Professor R. Narasimhan, to whom I express my thanks. The question was treated along other lines in [17].

Uniqueness. We show that if f is the constant zero on Γ, then f_1, f_2 and σ are the constant zero on their respective domains. The uniqueness of (f_1, f_2, σ) for given f then follows. From

$$f_1^*(s) + f_2^*(s) + \sigma(s) = 0$$

p.p. on Γ, we conclude with the aid of univalent conformal maps of an annulus $\{\rho < |z| < \rho^{-1}\}$ into S of the type described above and the Cauchy integral argument of Theorem 2, this Chapter, that if f_1 is the restriction of a member of $\mathcal{E}$ to Ω_1 which must be the constant zero since it is analytic at $\alpha(b)$ and takes the value 0 at a. Hence f_1 is the constant zero. We see that f_2 is the restriction to Ω_2 of $-\sigma$ which also is the constant zero and so f_2 and σ are the constant zero on their respective domains.

Existence. We recall that n is the common value of the valence of u. We suppose that f has domain Γ. We start the proof by observing that there exist uniquely determined functions $A_0, \ldots, A_{n-1}$ with domain $C(0;1)$ satisfying

$$f(s) = \sum_{0}^{n-1} A_k[u(s)][w(s)]^k, \tag{4.3}$$

$s \in \Gamma$. Further it is easily verified that $A_k \in L_p[C(0;1)]$, $k = 0, \ldots, n-1$ and that there exists a positive number c_1 such that for all allowed f we have

$$N[A_k \circ (u|\Gamma)] \leq c_1 N(f), \tag{4.4}$$

$k = 0, \ldots, n-1$.

Suppose that $F \in L_1[C(0;1)]$ and takes non-negative values. We seek to relate

$$\int_0^{2\Pi} F(e^{i\theta})d\theta$$

and

$$\int_{\Gamma} F \circ (u|\Gamma) \frac{\partial \mathcal{G}_a}{\partial n}\, ds.$$

We observe that because of the special conditions fulfilled by a unitary function we have

$$\log \frac{1}{(u)} = \sum_{1}^{n} \mathcal{G}_{\omega(k)}$$

where $\omega(1),\ldots,\omega(n)$ are points of Ω. There exist positive numbers c_2 and c_3 such that

$$c_2 \mathcal{G}_a(t) \leqslant \mathcal{G}_{\omega(k)}(t) \leqslant c_3 \mathcal{G}_a(t),$$

$k = 1,\ldots,n,$ for t in Ω near Γ. We note that

$$n\int_0^{2\Pi} F(e^{i\theta})\,d\theta = \int_{\Gamma} F \circ (u|\Gamma)\Sigma \frac{\partial \mathcal{G}_{\omega(k)}}{\partial n}\, ds ,$$

and hence conclude with the aid of the inequality of the preceding sentence that

$$c_2 \int F \circ (u|\Gamma)\frac{\partial \mathcal{G}_a}{\partial n}\, ds \leqslant \int_0^{2\Pi} F(e^{i\theta})\,d\theta$$

$$\leqslant c_3 \int_{\Gamma} F \circ (u|\Gamma)\frac{\partial \mathcal{G}_a}{\partial n}\, ds , \tag{4.5}$$

We are now in a position to obtain the desired representation for f. We let $A_{k,1}$ and $A_{k,2}$ denote the components of the M. Riesz decomposition of A_k. I.e., in the sense of Theorem 3, this Ch., with A_k taking the rôle of F, $A_{k,1}$ is the associated f and $A_{k,2}$ is the associated g. It suffices to prove the decomposition and the existence of a bound C for the case where $N(f) = 1$. We make this normalization. We introduce

$$\varphi_1 : s \to \Sigma A_{k,1}[u(s)][w(s)]^k, \; s \in \Omega,$$

and note that $\varphi_1 \in H_p(\Omega)$ and that there exists a positive number d_1 independent of allowed f such that $N(\varphi_1^*) \leq d_1$.

Continuing, we next introduce φ_2, the meromorphic function on $\alpha(\Omega)$ satisfying

$$\varphi_2(s) = \Sigma A_{k,2}[u(s)][w(s)]^k, \quad s \in \alpha(\Omega) - \{b\}.$$

The function φ_2 satisfies an inequality of the form

$$|\varphi_2(s)|^p \leq d_2 h(s)\exp[-p\nu \mathcal{G}_{\alpha(b)}(s)] \tag{4.6}$$

where d_2 is a positive number independent of allowed f, ν is $(n-1)$ times the order of the pole of w at b, and h is non-negative harmonic on $\alpha(\Omega)$ and satisfies $h[\alpha(a)] \leq 1$.

Finally we introduce $\sigma_o \in \mathfrak{S}, \beta \in \mathcal{B}$ such that

$$\lim_{s\to b} \varphi_2(s) - [\sigma_o(s) + \beta(s)] = 0.$$

It is to be observed that (σ_o,β) is thereby uniquely determined.

A representation of the desired kind is obtained by taking f_1 as $\varphi_1 + (\beta|\Omega) - \varphi_1(a)$, f_2 as $\psi - \psi[\alpha(a)]$ where $\psi = \varphi_2 - (\beta+\sigma_o)|\alpha(\Omega)$, and σ as $\sigma_o + \varphi_1(a) \qquad + \psi[\alpha(a)]$. We note that σ_o and β are bounded on Γ independently of allowed f. Indeed, thanks to (4.6) for each integer k the k th 0-Laurent coefficient of $\varphi_2 \circ \theta$, θ a uniformizer satisfying $\theta(0) = b$, is bounded independently of allowed f. On introducing a basis for $\mathfrak{S}$ whose members have divisors taking the values $0,1,\ldots,-g$ at b and a basis for $\mathfrak{B}$ whose members have poles at b of distinct orders $>g$, we see that the coefficients entering into the respective representations of σ_o and β in terms of the basis elements are bounded independently of allowed f. The asserted boundedness property of σ_o and β follows. It is now concluded with the aid of (4.6) that

$$|\psi|^p \leq H \tag{4.7}$$

where H is a non-negative harmonic function on $\alpha(\Omega)$ such that $H[\alpha(a)]$ is bounded independently of allowed f. Thus we are assured that each component of (f_1, f_2, σ) satisfies the conditions stated in (a) of the theorem.

There remains the question of the existence of C. We use the formula

$$M(|\varphi|^p)(t) = \frac{1}{2\Pi}\int_\Gamma |\varphi^*|^p \frac{\partial \mathcal{G}_t}{\partial n} ds, \quad t \in \Omega, \tag{4.8}$$

which is valid for $\varphi \in H_p(\Omega)$, $0 < p < +\infty$, and the corresponding one for functions in $H_p[\alpha(\Omega)]$. The formula (4.8) may be derived as follows. If q is a quasi-bounded non-negative harmonic function on Ω, with the aid of univalent conformal maps of the components of $\{|\mathcal{G}_t| < c\}$, c small and positive, onto plane annuli, using the mean convergence property of quasi-bounded harmonic functions on an annulus, we conclude from

$$q(t) = \frac{1}{2\Pi}\int_{\Gamma(\lambda)} q\frac{\partial \mathcal{G}_t}{\partial n} ds,$$

where $\Gamma(\lambda) = \{\mathcal{G}_t = \lambda\}$, λ small, on letting $\lambda \downarrow 0$, that

$$q(t) = \frac{1}{2\Pi}\int_\Gamma q^* \frac{\partial \mathcal{G}_t}{\partial n} ds.$$

We apply this equality to $q = M(|\varphi|^p)$ and use the observation that

$$\lim_{\lambda \to 0} \int_{\Gamma(\lambda)} [M(|\varphi|^p) - |\varphi|^p] \frac{\partial \mathcal{G}_a}{\partial n} ds = 0$$

from which we conclude that $M(|\varphi|^p)^* = |\varphi^*|^p$ p.p. (argument of Gårding and Hörmander cited in §6, Ch. II). The formula (4.8) follows.

Thanks to (4.8) we comclude that $|\varphi_1(a)| \leqslant N(\varphi_1^*) \leqslant d_1$ and thereupon that $N(f_1^*)$ is bounded above independently of allowed f. From (4.7) and the analogue of (4.8) for $\alpha(\Omega)$ we find that $N(f_2^*)$ is similarly bounded above. It is now immediate that the same is true for $N(\sigma|\Gamma)$. Theorem 5 now follows.

5. The theorem of Cauchy-Read. One of the important questions in the classical theory

of Hardy classes is the characterization of the Fatou boundary function of a member of $H_p(\Delta)$. This question was resolved for the case where $1 \leq p \leq +\infty$ by the classical work of F. and M. Riesz [31]. The corresponding problem for a compact bordered Riemann surface with regular analytic border was first treated by A.H. Read [30]. Another proof of the result of Read was given subsequently by H.Royden [33]. In this section we shall establish the results of Read with the aid of the provisional M.Riesz decomposition of Theorem 5 of the preceding section and a very useful Lemma of Havinson to be treated below. The theorem of Cauchy-Read may be stated as follows.

Theorem 6 (Read): Let $1 \leq p \leq +\infty$. The Fatou boundary function f^* of a function $f \in H_p(\Omega)$ belongs to $L_p(\Gamma)$ and satisfies

$$\int_\Gamma f^*\omega = 0 \tag{5.1}$$

whenever ω is an abelian differential (on some region of S) analytic at each point of $\bar{\Omega}$. In the opposite direction, if $F \in L_p(\Gamma)$ and (5.1) is satisfied with F replacing f^* for all allowed ω, then F is p.p. equal to the Fatou boundary function of a unique function $f \in H_p(\Omega)$. If, in particular, F is finite-valued and continuous, then $f \cup F$ is continuous.

Proof: That $f^* \in L_p(\Gamma)$ follows from the developments of the next to the last paragraph of the preceding section when $1 \leq p < +\infty$. The case $p = +\infty$ is obvious. We even have the "Poisson" representation

$$f(t) = \frac{1}{2\Pi}\int_\Gamma f^* \frac{\partial \mathcal{G}_t}{\partial n}\, ds, \quad s \in \Omega.$$

[cf. §5, Ch.II.] The uniqueness of f with a given Fatou boundary function follows. Using the Cauchy theorem and the mean convergence property of an H_1 function on an annulus we conclude the validity of (5.1) for all allowed ω. It is understood in (5.1) that Γ is positively sensed relative to Ω. Of course, the burden of the problem lies with the second assertion.

We first consider the case where $1 < p < +\infty$. For technical convenience we take b of Theorem 5 different from $\alpha(a)$. We introduce for $s \in S - \{\alpha(a)\}$, a function u_s harmonic on $S - \{s,\alpha(a)\}$, having a normalized positive logarithmic singularity at s and a normalized negative logarithmic singularity at $\alpha(a)$. We suppose that (f_1,f_2,σ) is the triple associated with F by Theorem 5. The first part of the present theorem yields

$$\int_\Gamma (f_2^* + \sigma)\delta u_s = 0, \quad s \in \alpha(\Omega) - \{\alpha(a)\}.$$

[For the operator δ, cf. §3, this Ch.] By the Cauchy theorem and the boundary behavior of f_2 we obtain

$$\int_\Gamma f_2^*\delta u_s = 2\Pi i f_2(s) = -\int_\Gamma \sigma\delta u_s$$

for the same admitted s. A second application of the Cauchy theorem shows that

$$\int_\Gamma \sigma\delta u_s = \int_\gamma \sigma\delta u_s$$

where γ is a small positively sensed contour surrounding $\alpha(b)$. We conclude that the function given by $\int_\Gamma \sigma\delta u_s$ admits harmonic prolongation and a posteriori analytic prolongation to $S - \{\alpha(b)\}$. The so obtained function, say φ, takes the value 0 at $\alpha(a)$ and has at worst a pole at $\alpha(b)$. The behavior of φ near $\alpha(b)$ may be concluded by noting the formula

$$\sigma(s) - \sigma[\alpha(a)] = \frac{1}{2\Pi i}\left[\int_\gamma + \int_{\gamma'} \sigma\delta u_s\right], \tag{5.2}$$

where γ' is a small positively sensed contour surrounding b, and the analyticity of

$$s \to \int_{\gamma'} \sigma\delta u_s$$

at $\alpha(b)$. The meromorphic extension of φ to S has a divisor whose value at $\alpha(b)$ is at least $-g$ and so φ is the constant 0. We see that f_2 is the constant 0. The representation (5.2) now shows that σ is analytic at $\alpha(b)$ and so σ is constant.

The asserted boundary property of F follows.

$p = \infty$. In this case the "Poisson" representation and the fact that, trivially, $F \in L_1(\Gamma)$, yield the desired result. In the special case where F is continuous, the asserted continuity of $f \cup F$ is concluded from the "Poisson" representation. Here the boundary behavior parallels that of the classical case of the Poisson integral.

There remains to be considered the case $p = 1$. It will be reduced to the case $p = +\infty$ with the aid of the following lemma of Havinson [2 8].

<u>Lemma</u> 3 (Havinson): <u>Let</u> $f \in L_1(\Gamma)$. <u>Let</u> (B_k) <u>be a sequence of functions analytic on</u> Ω, <u>each having modulus at most</u> 1, <u>which converges pointwise on</u> Ω <u>to</u> B. <u>Then</u>

$$\lim_{k\to\infty} \int_\Gamma fB_k^*\omega = \int_\Gamma fB^*\omega,$$

ω <u>being an analytic abelian differential on some open set containing</u> Γ.

The lemma follows simply by (1) noting its validity when f is analytic at each point of Γ thanks to the Cauchy integral theorem and the boundary behavior of the B_k and B, which permit us to replace the integrals along Γ by corresponding integrals along $\{\mathcal{g}_a = c\}$, c small and positive, and (2) thereupon obtaining the lemma in its full generality by approximating f in the mean of order 1 by a function analytic at each point of Γ, and making the appropriate estimates.

We shall not have occasion to use them, but we note that there are parallel results for sequences of $H_p(\Omega)$ functions of norm ≤ 1 which converge pointwise, the factor f being taken in $L_{p(p-1)^{-1}}(\Gamma)$, $1 < p < +\infty$.

We return to the proof of Theorem 6 and treat the remaining case: $p = 1$. It is to be observed that there exists φ analytic on Ω such that

$$\log|\varphi(t)| - \frac{1}{2\pi}\int_\Gamma \log(|F| + 1)\frac{\partial \mathcal{g}_t}{\partial n}\, ds \tag{5.3}$$

is bounded. Indeed, there exist 2g real harmonic functions on $S - \{b\}$(e.g. having singularities of the form $\mathrm{Re}(z^{-k})$, $\mathrm{Im}(z^{-k})$, $k = 1,\ldots,g$, in terms of suitable local uniformizers) such that the sum of

$$\frac{1}{2\pi}\int_\Gamma \log(|F| + 1)\frac{\partial g_t}{\partial n}\, ds$$

and a suitable linear combination of them (restricted to Ω) is the real part of an analytic function which is a logarithm of such a φ. The function $1/\varphi$ is bounded. There exists a sequence (B_k) of functions analytic at each point of $\bar{\Omega}$ such that $\max_{\bar{\Omega}} |B_k| = O(1)$ and (B_k) tends to $1/\varphi$ pointwise on Ω. This assertion may be established either by a Pick-Nevanlinna argument which would yield B_k of the form cu_k where c is a positive number independent of k and u_k is unitary (using interpolation to $1/\varphi$ on a suitable set of points) or else by appeal to the representation

$$\frac{1}{\varphi(t)} = \sum_0^{n-1} A_j[u(t)][w(t)]^j.$$

The A_j are bounded in a neighborhood of $C(0;1)$ and have poles at most at a finite number of points of Δ. We may approximate the A_j by rational functions R_j having the same principal parts in Δ as A_j and otherwise a pole at most at ∞ and indeed so that the approximating functions are uniformly bounded on $C(0;1)$. The resulting sequence of functions of the form

$$\sum_0^{n-1} R_j[u(t)][w(t)]^j$$

furnishes a sequence of the desired type.

By hypothesis we have

$$\int_\Gamma F(B_k\omega) = 0$$

for each B_k and each allowed ω. Hence by the Lemma of Havinson,

$$\int F\frac{1}{\varphi^*}\omega = 0.$$

Let β denote the bounded analytic function on Ω with Fatou boundary function equal to F/φ^* p.p. on Γ. Then the Fatou boundary function of $\varphi\beta$ is equal to F p.p.. Further $\varphi \in H_1(\Omega)$ as we see from the boundedness of (5.3) and an application of the inequality of the arithmetic and geometric means to the integral entering in (5.3). Hence $\varphi\beta \in H_1(\Omega)$. The proof of the theorem is complete.

6. $L_p(\Gamma)$ decomposition theorem (Final form). Thanks to the theorem of Cauchy-Read, it is possible to give a more satisfactory decomposition theorem for $L_p(\Gamma)$. This will be achieved by first determining the orthogonal complement of $H_2(\Omega)$ with respect to $L_2(\Gamma)$, the terms being appropriately interpreted, and by applying this information thereupon to $\sigma|\Gamma, \sigma \in \mathfrak{G}$. The terminal theorem will then be immediate.

Inner product. Given $F, G \in L_2(\Gamma)$ we introduce the inner product

$$\langle F,G\rangle = \frac{1}{2\Pi}\int_\Gamma F\bar{G}\frac{\partial \mathcal{G}_a}{\partial n}\,ds = \frac{i}{2\Pi}\int_\Gamma F\bar{G}\,\partial\mathcal{G}_a.$$

We "identify" $f \in H_2(\Omega)$ with f^* and propose the question of determining the orthogonal complement of $H_2(\Omega)$ so construed with respect to $L_2(\Gamma)$. To that end it will be convenient to introduce a notion generalizing that of a Hardy class lightly. We consider p, $0 < p < +\infty$, T a hyperbolic Riemann surface, and ∂ a divisor on T such that $\{\partial(t) \neq 0\}$ is finite. By the class $H_p(T,\partial)$ we understand the class of meromorphic functions f on T such that

$$\partial_f + \partial \geq 0,$$

and that for some compact $K \supset \{\partial(t) > 0\}$,

$$|f|(T - K)|^p$$

has a harmonic majorant. An equivalent definition is that $H_p(T,\partial)$ is the set of meromorphic functions f on T such that

$$|f|^p \exp[-p\Sigma\partial(t)\mathcal{G}_t]$$

have a harmonic majorant, $\mathcal{G}_t$ denoting for the moment Green's function for T with pole t. It is understood that the appropriate definitions are made at the points t where $\partial(t) \neq 0$. We return to the question under consideration and show

<u>Theorem</u> 7: <u>The orthogonal complement of</u> $H_2(\Omega)$ <u>with respect to</u> $L_2(\Gamma)$ <u>is</u>

$$H_2[\alpha(\Omega), \partial_{\delta \mathcal{G}_a} | \alpha(\Omega)]. \tag{6.1}$$

<u>Here the members of</u> (6.1) <u>are identified with their Fatou boundary functions as above.</u>

Proof: Given $f \in H_2(\Omega)$ and φ a member of (6.1), we see that

$$f\bar{\varphi} \circ \alpha\delta \mathcal{G}_a$$

is an analytic abelian differential on Ω. We find as a consequence of the boundary behavior of f and φ and the Cauchy integral theorem that $\langle f^*, \varphi^* \rangle = 0$, so that (6.1) is contained in the orthogonal complement of $H_2(\Omega)$ with respect to $L_2(\Gamma)$.

To proceed in the opposite direction we consider $\psi \in L_2(\Gamma)$ orthogonal to each member of $H_2(\Omega)$ and show that ψ "is a member of" (6.1). The theorem will follow. We note that there exists a zero-free analytic abelian differential on some open set containing $\bar{\Omega}$. Indeed, if $b \in S$, we are assured by the Weierstrass gap theorem that there exists a function analytic on $S - \{b\}$ which has a simple zero at an assigned point of $S - \{b\}$ but has no other zeros. We form by multiplication, using an abelian differential on S, not the zero differential, and integer powers of such analytic functions serving to compensate the zeros and poles of the abelian differential in $S - \{b\}$, an analytic abelian differential on $S - \{b\}$ free from zeros. We take $b \in \alpha(\Omega)$ and obtain a differential ω_o of the desired type. Thanks to the hypothesis on ψ,

$$\int_\Gamma \frac{\omega}{\omega_o} \bar{\psi} \delta \mathcal{G}_a = 0$$

for all ω allowed in the theorem of Cauchy-Read, or what is the same,

$$\int_\Gamma \bar{\psi} \frac{\partial \mathcal{G}_a}{\omega_o} \omega = 0.$$

By the Cauchy-Read theorem there exists a member of $H_2(\Omega)$ whose Fatou boundary function is equal to $\bar{\psi}\delta \mathcal{G}_a/\omega_o$ p.p. on Γ. We conclude that ψ is equal to the Fatou boundary function of a member of (6.1) p.p. on Γ. The theorem follows.

An application. Given $\sigma \in \mathfrak{S}$ (of §4, this Ch.) we have

$$\sigma(t) = f_1^*(t) + f_2^*(t) \tag{6.2}$$

p.p. on Γ for a unique $(f_1,f_2), f_1 \in H_2(\Omega)$ and f_2 in (6.1). It is to be noted that $H_2(\Omega)$ is a closed linear subspace of $L_2(\Gamma)$, appropriate conventions prevailing. Since σ is analytic at each point of Γ, it follows that f_1 and f_2 are both restrictions of functions analytic at each point of Γ. Let $p, 1 < p < +\infty$, be given. We observe that there exists a positive number C such that

$$N(f_1^*),\ N(f_2^*) \leqslant CN(\sigma) \tag{6.3}$$

for $\sigma \in \mathfrak{S}$, where N is the L_p norm defined by (4.1), this Ch.. This result is easily established with the aid of a basis for $\mathfrak{S}$. Indeed, if $\sigma_1,\ldots,\sigma_{g+1}$ are the members of such a basis, we have

$$\min N\Big(\sum_1^{g+1} c_k\sigma_k\Big) > 0$$

where the c_k are complex and $\max\limits_{1\leqslant k\leqslant g+1} |c_k| = 1$. From this inequality we see that the basis coefficients of $\sigma \in \mathfrak{S}$ are bounded in modulus by $dN(\sigma)$ where d is a positive number independent of σ. The inequality (6.3) follows on representing a member of $\mathfrak{S}$ as a linear combination of the basis elements and decomposing the basis elements according to (6.2). The decomposition theorem in its final form is now easily treated.

<u>Theorem</u> 8: <u>Let</u> $1 < p < +\infty$. (a) <u>Given</u> $F \in L_p(\Gamma)$, <u>there exists</u> (f_1,f_2) <u>unique,</u> <u>where</u> $f_1 \in H_p(\Omega)$, $f_2 \in H_p[\alpha(\Omega), \partial_\delta \mathcal{G}_a | \alpha(\Omega)]$, <u>satisfying</u>

$$F(t) = f_1^*(t) + f_2^*(t), \tag{6.4}$$

<u>p.p. on</u> Γ. (b) <u>There exists a positive number</u> C <u>such that for each</u> $F \in L_p(\Gamma)$ <u>the</u> <u>associated</u> (f_1,f_2) <u>satisfy</u>

$$N(f_1^*),\ N(f_2^*) \leq CN(F).$$

The proof is now simple. The existence part of (a) follows from Theorem 5 and (6.2), this Ch.. The uniqueness is referred to the L_2 situation. Thus when F is the zero constant on Γ, candidate f_1 and f_2 are restrictions of functions analytic at each point of Γ. We find that f_1^* is orthogonal to itself. Uniqueness follows. (b) follows from Theorem 5 and (6.3), this Ch..

Theorem 8 is justifiably considered as a final form of a decomposition theorem for $L_p(\Gamma)$ since the only arbitrary element entering is the normalization point a.

7. Linear functionals on $H_p(\Omega)$, $1 \leq p < +\infty$. Representation formulas for bounded linear linear functionals on $H_p(\Omega)$, $1 \leq p < +\infty$, may be obtained very rapidly with the aid of the Hahn-Banach extension theorem in complex form (Bohnenblust-Sobczyk [20]) - to be referred to as "HBBS" - and the classical F.Riesz representation for bounded linear functionals on $L_p(\Gamma)$. Thus if λ is a bounded linear functional on $H_p(\Omega)$, we introduce $\{(f^*, \lambda(f) : f \in H_p(\Omega)\}$, extend it by HBBS to $L_p(\Gamma)$ and conclude that there exists $\varphi \in L_{p/(p-1)}(\Gamma)$ such that

$$\lambda(f) = \int_\Gamma f^*\bar{\varphi}\delta\mathcal{G}_a, \tag{7.1}$$

$f \in H_p(\Omega)$.

When we consider the case where $1 < p < +\infty$, we may obtain without the intervention of HBBS but using the F.Riesz representation and Theorem 8 of the preceding section a representation theorem with uniqueness. Given $F \in L_p(\Gamma)$ we denote the first component f_1 of the decomposition of Theorem 8 by $\Pi(F)$. Π is a "projection". We see that $\lambda \circ \Pi$ is a bounded linear functional on $L_p(\Gamma)$. We have (7.1) again with some $\varphi \in L_{p/(p-1)}(\Gamma)$. Applying Theorem 8 to φ we introduce (φ_1, φ_2), the uniquely associated pair and conclude with the aid of the boundary behavior of the entering functions and the Cauchy integral theorem that

$$\lambda(f) = \int_\Gamma f^*\bar{\varphi}_1^*\delta\mathcal{G}_a, \tag{7.2}$$

$f \in H_p(\Omega)$. Further φ_1 is the unique member of $H_{p/(p-1)}(\Omega)$ having this property. For if $\psi \in H_{p/(p-1)}(\Omega)$ has the property that

$$\int_\Gamma f^* \bar{\psi}^* \delta \mathcal{G}_a = 0,$$

$f \in H_p(\Omega)$, we see on introducing ω_o of the preceding section that

$$\int_\Gamma \frac{\omega}{\omega_o} \bar{\psi}^* \delta \mathcal{G}_a = 0,$$

for all ω of the Cauchy-Read theorem. Hence ψ^* is the Fatou boundary function of a member of

$$H_{p/(p-1)}[\alpha(\Omega), \partial_{\delta \mathcal{G}_a} | \alpha(\Omega)]$$

and because of this fact we infer that ψ is the constant 0 by the uniqueness assertion of Th. 8, preceding §.

8. An approximation theorem for $H_p(\Omega)$, $1 < p < +\infty$. We consider the smallest closed linear subspace $\mathcal{L}$ of $H_p(\Omega)$ generated by the family of functions which are restrictions of unitary functions to Ω and show

Theorem 9: $\mathcal{L} = H_p(\Omega)$.

The proof is established by showing that a bounded linear functional on $H_p(\Omega)$ which vanishes on $\mathcal{L}$ vanishes identically. For then HBBS assures that $\mathcal{L} = H_p(\Omega)$. (Contrapositively, if $\mathcal{L} \neq H_p(\Omega)$, there exists a bounded linear functional vanishing on $\mathcal{L}$ but not identically on $H_p(\Omega)$.)

Suppose then that λ is a bounded linear functional on $H_p(\Omega)$ which vanishes on $\mathcal{L}$. We use the representation (7.1) of the preceding section. Since $\lambda(u) = 0$ whenever u is the restriction of a unitary function to Ω, the lemma of Havinson and the possibility of approximating an analytic function on Ω of modulus ≤ 1 boundedly by a sequence of such u yields: $\lambda(B) = 0$ whenever B is a bounded analytic function on Ω.

Suppose now that $f \in H_p(\Omega)$ but is not the constant zero. Introducing the term Q of (2.1), Ch.II, relative to $\log|f|$, we observe that by (2.2), Ch.II,

$$M|f|^p \geqslant \exp \circ (pQ) \geqslant |f|^p,$$

and consequently, Q^* is equal to $\log|f^*|$ p.p. on Γ. It follows that

$$Q(t) = \frac{i}{2\Pi}\int_\Gamma \log|f^*|\delta\mathcal{G}_t, \quad t \in \Omega. \tag{8.1}$$

Let b be a non-Weierstrass point of S lying in $\alpha(\Omega)$. Let v_k, w_k be real harmonic functions on $S - \{b\}$ and have respectively singularities at b given locally by $\mathrm{Re}(z^{-k})$ and $\mathrm{Im}(z^{-k})$ in terms of suitably chosen uniformizers, $k = 1,\ldots,g$. The choice of the v_k and w_k is not inherently compelling. What are wanted are functions u bounded, real and harmonic on Ω such that the period systems of the δu form a basis for the set of period systems of the δH, H real harmonic on Ω. There exist $u_1,\ldots,u_g$ drawn from the $v_k|\Omega$ and $w_k|\Omega$ having the desired property.

Let h be analytic on Ω and satisfy

$$\log|h| = Q + \sum_1^g c_k u_k,$$

where the real numbers c_k are uniquely specified by the requirement that $\log|h|$ be the real part of an analytic function, and in addition let $h(a) > 0$. The function h is thereby specified uniquely. For each whole number n the function h_n is analogously defined with $Q(t)$ being replaced by

$$\frac{i}{2\Pi}\int_\Gamma \min\{\log|f^*|,n\}\delta\mathcal{G}_t.$$

We see that f/h and each h_n is bounded and further (h_n/h) is a bounded sequence tending pointwise to 1. Referring to the last sentence of the second paragraph back, we see that $\lambda[(f/h)h_n] = 0$, $n = 0,1,\ldots$, or equivalently,

$$\int_\Gamma f^*\bar{q}\left(\frac{h_n}{h}\right)^* \delta\mathcal{G}_a = 0, \quad n = 0,1,\ldots .$$

By the lemma of Havinson $\lambda(f) = 0$. The theorem follows.

9. In this section we return to the "Toeplitzian" question studied in §7, Ch. II and obtain further theorems of the kind developed in that section but where the finite topological characteristics of the underlying Riemann surface as well as non-degeneracy enter into the argument. Here we take our surface to be Ω. The notations θ_f, Θ_F, ν are taken in the present context. $\| \ \|$ will refer to $H_p(\Omega)$ with normalization point a.

The first theorem to be proved appeals to the classical theorem concerning lower semi-continuous functions on a space of the second category which omit $+\infty$, the one that asserts that for such a function φ there exists a real number c such that $\text{int}\{\varphi(x) \leq c\} \neq \emptyset$.

<u>Theorem</u> 10: <u>Let</u> $1 \leq p < +\infty$. <u>Then</u> θ_f <u>maps</u> $H_p(\Omega)$ <u>into itself if and only if</u> f <u>is analytic on</u> Ω <u>and bounded</u>.

Proof: "if" is immediate. In the opposite sense we infer from the hypothesis that $f^n \in H_p(\Omega)$, $n = 0,1,\ldots$, so that

$$f \in \cap_{0<q\leq+\infty} H_q(\Omega).$$

This result is, of course, valid in the general situation of an unrestricted Riemann surface. We put aside the trivial case where f is the constant 0. By Th.2, Ch.II for $\varphi \in H_p(\Omega)$,

$$M|f\varphi|^p = M[\exp \circ (pQ)|\varphi|^p] \tag{9.1}$$

where Q is the term so designated in (2.1), Ch.II.,relative to $\log|f|$. Let (h_n) denote a non-decreasing sequence of harmonic functions on Ω, each bounded above, tending to Q. Let v_n denote the negative of a finite sum of minimal positive harmonic functions on Ω such that $h_n + v_n$ is the real part of an analytic function (cf. §3, this Ch.) and let ψ_n be an analytic function on Ω satisfying $\log|\psi_n| = h_n + v_n$. Now ψ_n is bounded and so $\psi_n\varphi \in H_p(\Omega)$. By Th.2, Ch.II.,we have

$$M(|\psi_n\varphi|^p) = M[\exp \circ (ph_n)|\varphi|^p]. \tag{9.2}$$

Given m a whole number,

$$\exp \circ (ph_m)|\varphi|^p \leqslant \lim_{n\to\infty} M[\exp \circ (ph_n)|\varphi|^p]$$

$$\leqslant M[\exp \circ (pQ)|\varphi|^p],$$

whence it follows that

$$\lim_{n\to\infty} M[\exp \circ (ph_n)|\varphi|^p] = M[\exp \circ (pQ)|\varphi|^p].$$

Since

$$(M[\exp \circ (ph_n)|\varphi|^p])$$

is a non-decreasing sequence, by (9.1) and (9.2) we see that $\|\theta_f(\varphi)\|$ is the limit of the monotone non-decreasing sequence $(\|\theta_{\psi n}(\varphi)\|)$. Now $\varphi \to \|\theta_{\psi_n}(\varphi)\|$ is continuous on $H_p(\Omega)$. Hence $\varphi \to \|\theta_f(\varphi)\|$ is lower semi-continuous on $H_p(\Omega)$. It is bounded on a ball by the cited classical result on lower semi-continuous functions. By properties of a norm, $\varphi \to \|\theta_f(\varphi)\|$ is bounded on the unit ball. For some positive number c, θ_{cf} maps the unit ball of $H_p(\Omega)$ into itself. It now follows from Th.7, Ch.II, that f is bounded.

A question that is connected with the theory of Toeplitz forms is the following. Let F be a bounded real-valued Lebesgue measurable function on Γ. Let $0 < p < +\infty$. We introduce

$$\mu_1 = \sup \frac{i}{2\Pi} \int_\Gamma F|\varphi^*|^p \delta \mathcal{G}_a,$$

where $\varphi \in H_p(\Omega)$ and $\|\varphi\| = 1$, and also

$$\mu_2 = \sup \frac{i}{2\Pi} \int_\Gamma F\, u^*\, \delta \mathcal{G}_a,$$

where u is PL on Ω and $\nu(u) = 1$. Then we have

<u>Theorem</u> 11: $\mu_1 = \mu_2 = \text{ess. sup. } F$.

We recall that the essential supremum of F is simply the minimum of the set of real c such that meas.$\{F(t) > c\} = 0$. The existence of a Fatou 'radial' limit of u p.p. on Γ is assured by the F.Riesz decomposition for subharmonic functions and Littlewood's radial limit theorem. [This assertion is to be construed in terms of uniformizers which map the unit circumference onto the components of Γ. The resulting u^* are equal to one another p.p.. The license taken in the x notation for u is harmless, though, to be sure, reference to sectorial limits is not warranted. For an allowed u we have

$$u^* = (Mu)^*$$

p.p. on Γ. Further, for an allowed u there exists an allowed φ such that $|\varphi^*|^p = u^*$ p.p. on Γ. Indeed, such a φ is obtained by introducing the term Q of (2.1), Ch. II, relative to $\log u$, taking v as the negative of a finite sum of minimal positive harmonic functions on Ω such that $Q + v$ is the real part of an analytic function, and thereupon taking φ satisfying $p \log|\varphi| = Q + v$. We conclude that $\mu_2 \leqslant \mu_1$. Since $|\varphi|^p$ is an allowed u for each allowed $\varphi, \mu_1 \leqslant \mu_2$. Consequently $\mu_1 = \mu_2$.

To determine the common value of μ_1 and μ_2 we consider the case where F has a positive lower bound; the general case follows by suitably translating F. Without invoking any restriction it is trivial to see that $\mu_2 \leqslant$ ess. sup F. We proceed in the other direction, taking advantage of the restriction to introduce

$$G(t) = \exp\left(\frac{i}{2\Pi} \int_{\Gamma} \log F \, \delta \mathcal{G}_t\right),$$

$t \in \Omega$. On applying Th. 6, Ch. II, to Θ_{G/μ_2}, we conclude that $G(t) \leqslant \mu_2$, $t \in \Omega$, and thence ess. sup $F \leqslant \mu_2$. The theorem follows.

Chapter V

Vector-Valued Functions

1. In this chapter we shall explore aspects of the theory of Hardy classes of vector-valued analytic functions. It is not a mere routine paraphrase of the classical theory. For one thing, even such standard phenomena as the p.p. existence of radial limits are no longer assured. Nevertheless, the boundary studies of Ch. IV may be extended to the vector-valued case when one replaces the boundary function by a suitably restricted vector-valued Radon measure on the boundary. (By a vector-valued Radon measure on the boundary we understand a bounded linear map of the space of complex-valued continuous functions on the boundary into the fixed Banach space under consideration.)

2. Basic definitions. We assume that X is a fixed Banach space over C. The norm in X will be denoted by $\| \ \|$. Let f mapping a neighborhood of a point a of a Riemann surface into X be given. The notions of the analyticity of f at a and the harmonicity of f at a are referred to the plane case by means of a local uniformizer. For the plane case analyticity will be defined in a customary way - via the existence of the derivate (in the strong sense) at each point of a neighborhood. The classical apparatus extends without the intoduction of any essentially new ideas. Continuing in the plane case we introduce the notion of harmonicity at $\alpha \in C$ by any one of the following equivalent means: (1) the existence of a local representation of the form $\varphi(z) + \psi(\bar{z})$ where φ is analytic at α and ψ at $\bar{\alpha}$, (2) the validity of the Laplace equation $u_{z\bar{z}} = 0$ in a neighborhood of α, (3) the validity of the Gauss-Koebe mean-value property, i. e., for each z sufficiently near α the function u under consideration (assumed continuous in a neighborhood of α) satisfies

$$u(z) = \frac{1}{2\Pi}\int_0^{2\Pi} u(z + re^{i\theta})d\theta$$

for sufficiently small positive r. We assume that the usual prelimaries concerning these matters are at our disposal. Their treatment is routine.

Using (3) we conclude that if f is harmonic (in particular analytic) on an open set O of a Riemann surface, then $\|f\|$ is subharmonic on O. When f is analytic on O we have the stronger conclusion that $\log\|f\|$ is subharmonic on O. To see this, it suffices to consider the plane case where f is analytic at 0 and to show that the mean-value property holds relative to 0. We introduce a linear functional λ on X of norm 1 which satisfies $\lambda[f(0)] = \|f(0)\|$. Using the subharmonicity of $\log|\lambda \circ f|$ in a neighborhood of 0 we infer that

$$\log\|f(0)\| \leq \frac{1}{2\Pi}\int_0^{2\Pi} \log|\lambda \circ f(re^{i\theta})|d\theta$$

$$\leq \frac{1}{2\Pi}\int_0^{2\Pi} \log\|f(re^{i\theta})\|d\theta$$

for small positive r. This is the customary proof. The following ingenious proof of the subharmonicity of $\log\|f\|$ for f analytic is due to D.S.Mitrinović. Let η be positive and let φ be a complex-valued analytic function on $\Delta(0;r)$, r small, satisfying

$$\log|\varphi(z)| = -\frac{1}{2\Pi}\int_0^{2\Pi} \log(\|f(re^{i\theta})\| + \eta)k(\theta,\frac{z}{r})d\theta,$$

$|z| < r$. Then $f(z)\varphi(z)$ defines an analytic function on $\Delta(0;r)$. Using the subharmonicity of $\|f(z)\||\varphi(z)|$ on $\Delta(0;r)$ we have from the maximum principle: $\|f(0)\||\varphi(0)| < 1$. Hence $\|f(0)\| < 1/|\varphi(0)|$. The proof follows on taking logarithms and thereupon letting η tend to 0.

The fact that $\log\|f\|$ is subharmonic for f analytic leads naturally to the following formulation of the notion of a Hardy class in the present setting. Given a Riemann surface T, f analytic on T taking values in X will be said to belong to the Hardy class $H_\infty(T;X)$ provided that f is bounded, to the Hardy class $H_p(T;X)$, $0 < p < +\infty$, provided the subharmonic function $\|f\|^p$ has a harmonic majorant.

3. Fatou limit theorems. The object of this section is to raise a question and to show by example what complications are to be expected. The question is this: For what complex Banach spaces X is it the case that the Fatou radial limit theorem holds

for a vector-valued analytic function f with values in X subject to the proviso that it satisfy one of the same restrictive conditions as those of the classical theory, e.g., that $\overset{+}{\log}\|f\|$ have a harmonic majorant? We shall examine the question for $X = l_p(C)$, $1 \leq p \leq +\infty$, and shall see that for $p = +\infty$, the Fatou property need not hold even for f that are bounded in norm, and that, in contrast, the Fatou property persists when $1 \leq p < +\infty$.

The case $p = +\infty$ is disposed of by a simple example. It will be assumed that the sequences entering throughout this section have as their domain the set of whole numbers. We consider the map $f : \Delta \to l_\infty(C)$ given by $f(z) = (z^k), |z| < 1$. It is readily checked that f is analytic and $\|f(z)\| = 1, |z| < 1$. Nevertheless, $\lim\limits_{r\to 1} f(r\zeta)$ does not exist for any $\zeta \in C(0;1)$ as is easily verified.

We turn to the positive theorem.

<u>Theorem</u> 1: <u>Given</u> $1 \leq p < +\infty$. <u>Let</u> $f : \Delta \to l_p(C)$ <u>be analytic on</u> Δ <u>and such that</u> $\overset{+}{\log}\|f\|$ <u>have a harmonic majorant. Then</u> f <u>has a Stoltz limit</u> p.p. <u>on</u> $C(0;1)$.

<u>Corollary</u>: Theorem 1 <u>persists when</u> Δ <u>is replaced by</u> Ω of Ch. IV.

Proof: We introduce φ, a complex-valued analytic function on Δ satisfying $\log|\varphi| = M\,\overset{+}{\log}\|f\|$. Replacing f by $f\varphi^{-1}$ we obtain a vector-valued analytic function taking values of norm ≤ 1. Once Theorem 1 is demonstrated for allowed functions taking values of norm ≤ 1, it follows unrestrictedly on noting that φ is the reciprocal of a non-vanishing bounded analytic function on Δ and hence has a finite Stoltz limit p.p. on $C(0;1)$.

We turn to the restricted case where f takes values of norm ≤ 1. We let $f_k(z)$ denote the kth component of $f(z), |z| < 1, k = 0,1,\dots$. Each f_k is analytic on Δ and $\Sigma|f_k|^p$ is subharmonic on Δ. We introduce $h_k = M|f_k|^p, k = 0,1,\dots$. It is given by the Poisson-Lebesgue integral with boundary function $|f_k^*|^p$. We also introduce $h = M(\Sigma|f_k|^p)$ and note that it is given by the Poisson-Lebesgue integral with boundary function $s = \Sigma|f_k^*|^p$. We let F denote the map of $[-\Pi,\Pi]$ into the set of complex-valued sequences satisfying the condition that the k th component of $F(\theta)$ is simply

$$\int_{-\Pi}^{\theta} f_k^*(e^{i\alpha})\,d\alpha.$$

It is readily verified with the aid of the Hölder inequality that F is a Lipschitzian map of $[-\Pi,\Pi]$ into $l_p(C)$. Further

$$f(z) = \frac{1}{2\Pi}\int_{-\Pi}^{\Pi} k(\theta,z)\,dF(\theta), \quad |z| < 1.$$

Paraphrasing the proof of the classical Fatou theorem for functions representable by a Poisson-Stieltjes integral [16] we see that the radial limit of f exists at $e^{i\theta}$ when $F'(\theta)$ exists. The existence of a Stoltz limit at a point $e^{i\theta}$ for which a radial limit exists is concluded by examining the behavior of the subharmonic function $\log\|f - f^*(e^{i\theta})\|$, which is bounded above, with the aid of standard arguments employed for the same purpose in the classical case. cf. [16,p. 105].

Thus our problem is reduced to exhibiting $E \subset [-\Pi,\Pi]$ and having Lebesgue measure 2Π at each point of which F' exists. We assert that an admissible choice of E is the subset of $[-\Pi,\Pi]$ at each point θ of which the following conditions are fulfilled: (1) each of the f_k^* is defined at $e^{i\theta}$ and $f_k^*(e^{i\theta})$ is the derivative at θ of the k th component of F, (2) the derivatives of the functions

$$\varphi \to \int_{-\Pi}^{\varphi} |f_k^*(e^{i\alpha})|^p d\alpha, \quad k = 0,1,\ldots,$$

and

$$\varphi \to \int_{-\Pi}^{\varphi} s(e^{i\alpha})\,d\alpha$$

exist and are respectively equal to $|f_k^*(e^{i\theta})|^p$, $k = 0,1,\ldots,$ and $s(e^{i\theta})$. [Reference is made to s introduced in the preceding paragraph.] Indeed, the set so defined has Lebesgue measure 2Π. Further given $\theta \in E$ we see that for $\varphi \in [-\Pi,\Pi]$ different from θ the p th power of the norm of

$$[F(\varphi) - F(\theta)](\varphi-\theta)^{-1} - (f_k^*(e^{i\theta}))$$

does not exceed the sum of

$$\sum_{0}^{m} \left| \frac{\int_{\theta}^{\varphi} f_k^*(e^{i\alpha})\,d\alpha}{\varphi - \theta} - f_k^*(e^{i\theta}) \right|^p$$

and

$$2^{p-1}\left[\sum_{m+1}^{\infty} \left| \frac{\int_{\theta}^{\varphi} |f_k^*(e^{i\alpha})|^p d\alpha}{\varphi - \theta} \right| + \sum_{m+1}^{\infty} |f_k^*(e^{i\theta})|^p\right] \tag{3.1}$$

for each whole number m. Now

$$\sum_{m+1}^{\infty} \left| \frac{\int_{\theta}^{\varphi} |f_k^*(e^{i\alpha})|^p d\alpha}{\varphi - \theta} \right| = \left| \frac{\int_{\theta}^{\varphi} [s(e^{i\alpha}) - \sum_{0}^{m} |f_k^*(e^{i\alpha})|^p] d\alpha}{\varphi - \theta} \right| .$$

Hence the limit at θ of (3.1) is equal to

$$2^p \sum_{m+1}^{\infty} |f_k^*(e^{i\theta})|^p.$$

Given the restriction imposed on the points of E and the arbitrariness of m, we conclude that $F'(\theta)$ exists and is equal to $(f_k^*(e^{i\theta}))$. The proof of Theorem 1 is complete.

Proof of Corollary. It suffices to map conformally and univalently a plane annulus into Ω so that one of boundary circumferences corresponds to a given component of Γ and to note that the boundary behavior of f is thereby referred to the case where the domain is an annulus, which case may be referred to the case of a disk with the aid of the Laurent representation. The details are readily furnished.

4. Vector-valued harmonic functions with norm possessing a harmonic majorant. From this point on we shall be operating in the setting of Ch. IV. For the present we consider functions u with domain Ω taking values in a complex Banach space which are harmonic and satisfy the condition that $\|u\|$ (resp. $\varphi \circ \|u\|$, where φ satisfies the conditions given in §1, Ch. II, has a harmonic majorant. In this section we shall see that the

functions in question are precisely those given by Poisson-Radon representations involving vector-valued boundary Radon measures subject to conditions to be stated below and that the Poisson-Radon representation defines a (1.1) map from the entering set of vector-valued Radon measures onto the set of harmonic functions coming into consideration. The state of affairs parallels that of the classical case ($\Omega = \Delta, X = C$) exactly. (F.Riesz-Herglotz representation).

The exposition that follows will be seen to be related to that which I gave in [16] for the proof of the classical F.Riesz-Herglotz theorem for non-negative harmonic functions. An element not present in the treatment of the classical situation is an approximation lemma (Lemma 2) which transplants to the situation we are studying an approximation property (Lemma 1) of the classical Poisson kernel.

Given $z \in \Delta$. We define K_z as

$$\eta \to \frac{1 - |z|^2}{|\eta - z|^2}, \quad |\eta| = 1.$$

<u>Lemma</u> 1: <u>Given</u> $0 \leqslant r < 1$. <u>Then</u> $\{K_z : r \leqslant |z| < 1\}$ <u>generates a dense linear subspace of the space of complex-valued continuous functions on</u> $C(0;1)$ (<u>with standard topology</u>). <u>In fact, a given continuous complex-valued function</u> f <u>on</u> $C(0;1)$ <u>may be approximated uniformly by a finite sum of the form</u> $\Sigma A_j K_{z_j}$ <u>where the</u> A_j <u>are complex numbers satisfying</u> $\Sigma |A_j| \leqslant \max|f|$ <u>and</u> $r \leqslant |z_j| < 1$, <u>all</u> j.

Proof: Using the Poisson integral for the unit disk we see that if f is a given complex-valued continuous function on $C(0;1)$ and u is the solution of the Dirichlet problem for Δ with boundary function f, then f, which may be approximated uniformly by $\eta \to u(\varrho\eta)$ for ϱ less then but sufficiently near 1, may be approximated uniformly by a function of the form

$$\eta \to \sum_{j=1}^{n} A_j K_{\varrho\eta}(\zeta_j) = \sum_{j=1}^{n} A_j K_{\varrho\zeta_j}(\eta), \tag{4.1}$$

where $\Sigma |A_j| \leqslant \max|f|$, for ϱ as above.

It is to be observed that the restriction on the coefficients will be of service immediately.

We next observe that a corresponding result holds for an annulus. We consider $\mathfrak{A} = \{r_0 < |z| < 1\}$, $0 < r_0 < 1$. We let $\mathfrak{G}_z$ denote Green's function for $\mathfrak{A}$ with pole z and let K_z now denote the function with domain $C(0;1)$ assigning to each point of $C(0;1)$ the inner normal derivative at that point. We assert that the conclusion of Lemma 1 persists for the present setting when $r_0 < r < 1$. To see this we introduce $\mathfrak{G}_z^0$, Green's function for Δ with pole $z(\in \mathfrak{A})$, and note that $\mathfrak{G}_z^0 | \mathfrak{A} = \mathfrak{G}_z + h_z$ where h_z is positive harmonic on $\mathfrak{A}$ and h_z tends uniformly to 0 as $|z| \to 1$. Consequently the difference of the present K_z and that of Lemma 1 tends to 0 uniformly as $|z| \to 1$. Our assertion now follows with the aid of the restriction imposed on the A_j in (4.1) as we see on replacing the 'K' of Lemma 1 by the corresponding 'K' pertaining to $\mathfrak{A}$.

We are now ready to turn to the desired lemma for Ω. We return to the standard notation for the Ω setting.

Let $\Gamma_1, \ldots, \Gamma_n$ be a univalent enumeration of the components of Γ. We fix a positive number C so that there are no zeros $\delta \mathcal{G}_a$ in $\{0 < \mathcal{G}_a(s) \leqslant C\}$ and the set $\{0 < \mathcal{G}_a(s) < C\}$ has n components, each an annulus. Each component of Γ is a component of the frontier of exactly one such annulus, and the frontier of each such annulus has a component which is a component of Γ. We index the annuli, $\mathfrak{A}_1, \ldots, \mathfrak{A}_n$ so that Γ_k is a component of $\mathrm{fr}\,\mathfrak{A}_k$. We introduce φ_k, a univalent conformal map of $\{\lambda_k < |z| < 1\}$, $0 < \lambda_k < 1$, onto $\mathfrak{A}_k$, where φ_k is so chosen that under its continuous extension to $\{\lambda_k \leqslant |z| \leqslant 1\}$, the unit circumference is mapped onto Γ_k, $k = 1, \ldots, n$. The level lines of $\mathcal{G}_a \circ \varphi_k$ are the circumferences $C(0;r), \lambda_k < r < 1$. We introduce

$$h_s = \frac{\delta \mathcal{G}_s}{\delta \mathcal{G}_a} \Big| \Gamma, \tag{4.2}$$

for each $s \in \Omega$ and show

<u>Lemma</u> 2: <u>Given</u> c <u>positive. The set</u>

$$\{h_s : 0 < \mathcal{G}_a(s) \leqslant c\}$$

<u>generates a dense linear subspace of the space of complex-valued continuous functions on</u> Γ (<u>with standard topology</u>).

Proof: The lemma follows once it is shown that for $j = 1,\ldots,\eta$ a complex-valued continuous function on Γ which vanishes on all components of Γ, save possibly the jth, may be approximated uniformly by a linear combination of the allowed h_s.

We note that each h_s is positive. Further h_s tends to 0 uniformly on $\Gamma - \Gamma_j$ as s tends to Γ_j (i.e. given $\varepsilon > 0$, there exists a neighborhood V of Γ_j such that $h_s(t) < \varepsilon$, $t \in \Gamma - \Gamma_j$, for $s \in V$). Indeed, $\mathcal{G}_s$ tends to 0 uniformly on a given compact subset of Ω as s tends to Γ, as may be seen with the aid of the Harnack inequality (2.1) of Ch.I, and the symmetry and boundary properties of a Green function for Ω. We consider $\varphi_j[C(0;r)]$, $\lambda_j < r < 1$, which is a compact subset of Ω, and introduce the harmonic measure ω of $\varphi_j[C(0;r)]$ with respect to $\Omega - \varphi_j(\{r \leqslant |z| < 1\})$. For $s \in \varphi_j(\{r < |z| < 1\})$, we have the inequality

$$h_s(t) \leqslant \left\{\max_{\varphi_j[C(0;r)]} \mathcal{G}_s\right\} \frac{\delta\omega(t)}{\delta\mathcal{G}_a}, \quad t \in \Gamma - \Gamma_j.$$

The asserted limit behavior of h_s follows.

We now consider the annulus $\{\lambda_j < |z| < 1\}$ and let $\mathfrak{G}_z$ and K_z be taken in the sense of this annulus. We see that

$$\mathcal{G}_{\varphi_j(z)} \circ \varphi_j - \mathfrak{G}_z \tag{4.3}$$

tends uniformly to zero as $|z| \to 1$. On taking the normal derivative of (4.3) at the points of $C(0;1)$ we conclude that

$$\frac{c}{\log(1/\lambda_j)}\, h_{\varphi_j(z)} \circ \varphi_j^* - K_z,$$

φ_j^* being the limit function of φ_j on $C(0;1)$, tends uniformly to 0 as $|z| \to 1$.

On combining the results of the last two paragraphs and the extension of Lemma 1 to an annulus, we conclude that a complex-valued continuous function on Γ which vanishes on all components of Γ save possibly the jth, may be approximated uniformly by a linear combination of the allowed h_s. The proof proceeds by exchanging K_z and

$$\frac{C}{\log(1/\lambda_j)} h_{\varphi_j(z)} \circ \varphi_j^*$$

and observing the effect on an approximation to $f \circ \varphi_k^*$. Lemma 2 then follows.

The role of Lemma 2 is to permit us to obtain a uniqueness lemma for vector-valued continuous linear maps from $C(\Gamma)$, the space of complex-valued continuous functions on Γ, (standard topology) into X (possibly C). Indeed, if λ is such a map, we have

<u>Lemma</u> 3: <u>If</u> $\lambda(h_s) = 0$ <u>when</u> $0 \leqslant g_a(s) \leqslant c$, <u>then</u> λ <u>is the null map. Here</u> c <u>is a positive number</u>.

The proof is immediate thanks to Lemma 2.

We now turn to the study of vector-valued functions on Ω constructed with the aid of λ as defined immediately before Lemma 3 and the family of functions h_s. We note that $(s,t) \to h_s(t)$ is continuous on $\Omega \times \Gamma$ and that $s \to h_s(t)$ is harmonic on Ω for each $t \in \Gamma$. Given an allowed λ we introduce H_λ with domain Ω defined by

$$H_\lambda(s) = \lambda(h_s).$$

It is readily verified that H_λ is harmonic. It follows from Lemma 3 that

$$\lambda \to H_\lambda$$

is univalent. We seek to characterize λ having the property that $\|H_\lambda\|$, resp. $\varphi \circ \|H_\lambda\|$ has a harmonic majorant, φ satisfying the conditions given in §1, Ch.II.

The central result is the following theorem.

Theorem 2: (i) A necessary and sufficient condition that $\|H_\lambda\|$ possess a harmonic majorant is that there exist a non-negative Radon measure μ on $C(\Gamma)$ such that

$$\|\lambda(f)\| \leq \mu(|f|), \tag{4.4}$$

for $f \in C(\Gamma)$. Every vector-valued harmonic function u on Ω with values in X such that $\|u\|$ admits a harmonic majorant is H_λ for a λ (necessarily unique) satisfying the condition (4.4) for some μ.

(ii) The corresponding result holds when $\varphi \circ \|\ldots\|$ replaces $\|\ldots\|$ in the text and (4.4) is replaced by the condition

$$\|\lambda(f)\| \leq \frac{1}{2\Pi}\int_\Gamma |f|\alpha\delta\sigma_a, \tag{4.5}$$

where α is a non-negative finite-valued Lebesgue measurable function on Γ such that $\varphi \circ \alpha$ is summable.

To be exact, we understand by a non-negative Radon measure on Γ a continuous linear map of $C(\Gamma)$ into $\mathbb{C}$ which takes each non-negative member of $C(\Gamma)$ into a non-negative real.

Proof of Theorem 2: The sufficiency of (4.4) (resp. (4.5)) is immediate on noting the positivity of h_s and for (4.5) the availability of the Jensen inequality.

The remainder of the proof proceeds along the lines of the proof which we have given for the F.Riesz-Herglotz representation theorem for non-negative harmonic functions [16].

We shall want the counterpart of some standard facts concerning Fourier coefficients associated with a complex-valued harmonic function on an annulus.

Lemma 4: Let h be a vector-valued harmonic function on $\{r_1 < |z| < r_2\}$, $0 \leq r_1 < r_2 \leq +\infty$, which takes values in X. Let $c_n(r)$ denote the nth Fourier coefficient of $h(re^{i\theta})$, $r_1 < r < r_2$, i.e.

$$c_n(r) = \frac{1}{2\Pi}\int_0^{2\Pi} h(re^{i\theta})e^{-ni\theta}d\theta.$$

Then $c_0(r) = A_0 \log r + B_0$, $c_n(r) = A_n r^n + B_{-n} r^{-n}$ when $n \neq 0$, where the coefficients A_n and B_n are elements of X. The coefficients are uniquely determined.

The lemma is an immediate consequence of the fact that h admits a representation of the form

$$h(z) = A \log|z| + f(z) + g(\bar{z}),$$

where $A \in X$ and f and g are vector-valued analytic functions on the annulus which take values in X. A proof may be given without the intervention of function-theoretic apparatus with the aid of the Weierstrass approximation theorem for vector-valued continuous functions and the maximum principle for subharmonic functions. cf. [19, p. 300 - 1].

We shall find it convenient to introduce a map of the level set $\Gamma(c) = \{\mathcal{G}_a(s) = c\}$ c small and positive,onto Γ. Returning to the φ_k introduced before Lemma 2 of this section, we observe that

$$\mathcal{G}_a \circ \varphi_k(z) = \frac{-\omega_k}{2\Pi} \log|z|, \quad \lambda_k < |z| < 1,$$

where

$$\omega_k = \int_{\Gamma_k} \frac{\partial \mathcal{G}_a}{\partial n} ds,$$

reference being made to the inner normal derivative. The map ψ of $\Gamma(c)$ onto Γ which we propose is

$$\cup_{1 \leq k \leq n}\{(\varphi_k(z), \varphi_k^*(\operatorname{sgz})) : |z| = e^{-\frac{2\Pi}{\omega_k}c}\}.$$

Further we shall agree that given a function f with domain Γ we shall denote $f \circ \psi$ by f^c.

To continue, we consider $u : \Omega \to X$, harmonic on Ω and such that $\|u\|$ has a harmonic majorant. Let $U = M\|u\|$. We introduce for c small and positive

$$\lambda_c(f) = \frac{i}{2\Pi} \int_{\Gamma(c)} u f^c \delta \mathcal{G}_a$$

and

$$\mu_c(f) = \frac{i}{2\Pi} \int_{\Gamma(c)} U f^c \delta \mathcal{G}_a,$$

$f \in C(\Gamma)$. It is understood that $\Gamma(c)$ has the standard positive sensing. The following inequality holds:

$$\|\lambda_c(f)\|, |\mu_c(f)| \leqslant \mu_c(|f|) \leqslant (\max_\Gamma |f|) U(a). \tag{4.6}$$

With the aid of the Weierstrass approximation theorem (trigonometric form) applied to $f \circ \varphi_k^*(e^{i\theta})$ and Lemma 4 of this section, we see that f may be approximated uniformly by a function $F \in C(\Gamma)$ which is such that $\lim_{c\to 0} \lambda_c(F)$ and $\lim_{c\to 0} \mu_c(F)$ exist. For c and d small and positive we have

$$\|\lambda_c(f) - \lambda_d(f)\| \leqslant \|\lambda_c(f - F)\| + \|\lambda_c(F) - \lambda_d(F)\| + \|\lambda_d(F - f)\|$$

and a corresponding inequality for $|\mu_c(f) - \mu_d(f)|$. We conclude that $\lim_{c\to 0} \lambda_c(f)$ and $\lim_{c\to 0} \mu_c(f)$ exist. We denote the former limit by $\lambda(f)$ and the latter by $\mu(f)$. The inequality (4.6) holds with λ replacing λ_c and μ replacing μ_c. It is now readily concluded that λ is a continuous linear map of $C(\Gamma)$ into X and that μ is a non-negative Radon measure on Γ.

We show that the following representations hold:

$$u = H_\lambda, \; U = H_\mu, \tag{4.7}$$

whence we conclude the second assertion of Theorem 2 (i), and thereupon the necessity of the first assertion using the univalence of $\lambda \to H_\lambda$. To establish (4.7) we introduce for c small and positive $\Omega_c = \{\mathcal{G}_a(s) > c\}$ and thereupon G_s^c, $s \in \Omega_c$, with domain a neighborhood of $\bar{\Omega}_c$ satisfying : (1) its restriction to Ω_c is Green's function for Ω_c with pole s, (2) G_s^c is harmonic at each point of Γ_c. The generalized Poisson integral formula

$$u(s) = \frac{i}{2\pi} \int_{\Gamma(c)} u \frac{\delta G_s^c}{\delta G_a^c} \delta \mathcal{g}_a, \quad s \in \Omega_c,$$ (4.8)

and the corresponding one for U hold. Now (4.8) may be rewritten as

$$u(s) = \lambda_c \left(\frac{\delta G_s^c}{\delta G_a^c} \circ \psi^{-1} \right),$$ (4.9)

a corresponding formula holding for U.

We shall use the fact that

$$\frac{\delta G_s^c}{\delta G_a^c} \circ \psi^{-1}$$ (4.10)

tends uniformly to h_s as c tends to 0. To see this, for $s \in \Omega$ given we fix a positive number d so that $2d < \mathcal{g}_a(s)$ and

$$\Gamma(c) \subset \bigcup_{1 \le k \le n} \lambda_k$$

when $0 < c \le 2d$. We note that with the aid of Schwarzian reflexion across $\Gamma(c)$ we are assured that G_s^c may be taken to have domain $\{\mathcal{g}_a(t) > -d\}$ and to be harmonic at each point of its domain save s provided that $0 < c < d/2$. We assume that c is so restricted and that G_s^c is so taken. Now the family of such G_s^c is bounded on

$$\{|\mathcal{g}_a(t)| < d\}.$$ (4.11)

It follows, thanks to the boundedness property, that G_s^c tends uniformly to $\mathcal{g}_s$ in (4.11) as c tends to 0. Consequently $\delta G_s^c / \delta G_a^c$ tends uniformly to $\delta \mathcal{g}_s / \delta \mathcal{g}_a$ in (4.11). Our assertion concerning (4.10) follows with the aid of the equality

$$\frac{\delta G_s^c}{\delta G_a^c} \circ \psi^{-1} - h_s = \left(\frac{\delta G_s^c}{\delta G_a^c} \circ \psi^{-1} - \frac{\delta \mathcal{g}_s}{\delta \mathcal{g}_a} \circ \psi^{-1} \right)$$

$$+ \left(\frac{\delta \mathcal{g}_s}{\delta \mathcal{g}_a} \circ \psi^{-1} - h_s \right).$$

Given $f_1, f_2 \in C(\Gamma)$ we see that

$$\|\lambda_c(f_1) - \lambda(f_2)\| \leqslant \|\lambda_c(f_1 - f_2)\| + \|\lambda_c(f_2) - \lambda(f_2)\|$$

$$\leqslant (\max_\Gamma |f_1 - f_2|) U(a) + \|\lambda_c(f_2) - \lambda(f_2)\|.$$

Taking into account this inequality, with (4.10) playing the role of f_1 and h_s that of f_2, we see that $u = H_\lambda$. The corresponding formula for U is concluded by the same argument.

The proof of Theorem 2(i) is now complete.

A minimal property of μ. Suppose that $\|H_\lambda\|$ possesses a harmonic majorant, that μ is the Radon measure obtained in the above proof, and that P is a non-negative harmonic function on Ω satisfying $P \geqslant \|H_\lambda\|$. With ν the non-negative Radon measure on $C(\Gamma)$ satisfying $P = H_\nu$, we have $H_{\nu-\mu} = H_\nu - H_\mu \geqslant 0$, whence $\nu - \mu$ is a non-negative Radon measure by the unicity of the representation $H_{\nu-\mu}$.

The non-negative Radon measure μ may be obtained by a simple limit process from λ. We introduce ν_c on $C(\Gamma)$ by

$$\nu_c(f) = \frac{i}{2\Pi} \int_{\Gamma(c)} \|H_\lambda\| f^c \delta \mathcal{g}_a.$$

Now

$$|\mu_c(f) - \nu_c(f)| \leqslant (\max_\Gamma |f|)[H_\mu(a) - \frac{i}{2\Pi} \int_{\Gamma(c)} \|H_\lambda\| \delta \mathcal{g}_a],$$

$f \in C(\Gamma)$, and the right side tends to 0 as c tends to 0. Hence ν_c tends to μ pointwise on $C(\Gamma)$.

The condition (4.4). We remark that this condition may be genuinely restrictive. It suffices to consider the case where $X = l_\infty(\mathbb{C})$. Suppose that $\lambda(f) = (f(t_k))$ where (t_k) is a fixed univalent sequence of points of Γ. Then λ is, of course, a continuous linear map of $C(\Gamma)$ into $l_\infty(\mathbb{C})$. But there does not exist a μ such that (4.4) is satisfied for all $f \in C(\Gamma)$.

Proof of Theorem 2(ii). When $\varphi \circ \|H_\lambda\|$ has a harmonic majorant, then by Th. 2, Ch.II, $H_\mu = M\|H_\lambda\|$ is quasibounded as is $M(\varphi \circ \|H_\lambda\|) = M(\varphi \circ H_\mu)$. The measure μ is given by

$$\mu(f) = \frac{i}{2\pi} \int_\Gamma H_\mu^* f \delta \mathcal{G}_a .$$

Indee, the non-negative Radon measure ν given by the right-hand side satisfies $H_\nu = H_\mu$. Further $\varphi \circ H_\mu^*$ is summable and is equal p.p. on Γ to the Fatou boundary function of $M(\varphi \circ H_\mu)$. It is now easy to establish (ii) of Theorem 2 on noting that H_μ^* serves for α in the necessity statement.

5. Boundary theory for $\mathbb{H}_p(\Omega,X)$, $1 \leq p \leq +\infty$. If $f \in \mathbb{H}_p(\Omega,X)$, $1 \leq p \leq +\infty$, then $f = H_\lambda$ for a continuous linear map λ of $C(\Gamma)$ into X, which satisfies (4.4), this Ch., for some non-negative Radon measure μ on $C(\Gamma)$. We are concerned in this section with characterizing the continuous linear maps λ of $C(\Gamma)$ into X such that $H_\lambda \in \mathbb{H}_p(\Omega,X)$.

Necessary conditions. When $p = +\infty$, we see that $M\|H_\lambda\|$ is bounded and so H_μ^* is bounded. Throughout this section it is understood that $H_\mu = M\|H_\lambda\|$. Suppose now that $1 \leq p < +\infty$. By Th. 2, Ch.II, with $\varphi(x) = \exp(px)$ and $u = \log\|H_\lambda\|$ we conclude that $M\|H_\lambda\|^p$ is quasi-bounded. We obtain from these observations a first necessary condition on λ with the aid of Th. 2(ii), this Ch.:

$$\text{(I)} \quad \|\lambda(f)\| \leq \frac{i}{2\pi} \int_\Gamma |f| \alpha \delta \mathcal{G}_a , \qquad (5.1)$$

$f \in C(\Gamma)$, where α is a non-negative member of $L_p(\Gamma)$.

A second necessary condition is obtained from the Cauchy theorem and limit considerations. Let ω be an analytic abelian differential on some open set containing $\bar{\Omega}$. With the notation of the preceding section in force with $u = H_\lambda$, allowing a slight notational license in writing $\omega/\delta\mathcal{g}_a$, we have

$$\lambda_c\left(\frac{\omega}{\delta\mathcal{g}_a}\circ\psi^{-1}\right) = \frac{i}{2\Pi}\int_{\Gamma(c)} H_\lambda \frac{\omega}{\delta\mathcal{g}_a}\,\delta\mathcal{g}_a = 0$$

for small positive c by the Cauchy integral theorem. On letting c tend to 0 we obtain

$$\text{(II)} \qquad \lambda\left(\frac{\omega}{\delta\mathcal{g}_a}\right) = 0 \tag{5.2}$$

for each admitted ω.

We now show that if λ satisfies (I) and (II), then $H_\lambda \in \mathbb{H}_p(\Omega,X)$. We obtain thereby a theorem of Cauchy-Read type for vector-valued functions.

It is clear from (I) that when $p = \infty$, then H_λ is bounded and when $1 < p < +\infty$, then $\|H_\lambda\|^p$ has a harmonic majorant.

There remains to be shown that H_λ is analytic on Ω. We accomplish this by showing that for each bounded linear functional l on X the function $l \circ H_\lambda$ is analytic. For then on representing H_λ locally in terms of a uniformizer as the sum of an analytic function and an analytic function composed with conjugation, we conclude that l composed with the latter function is constant (being analytic and antianalytic at the same time) and hence that the latter function is constant by the arbitrariness of l. The analyticity of H_λ follows.

To show the analyticity of $l \circ H_\lambda$ we proceed as follows. Let $\mathcal{L} = l \circ \lambda$. Then $\mathcal{L}$ satisfies (I) and (II), replacing λ, α being appropriately changed. From (I) we conclude that $\mathcal{L}$ admits a representation of the form

$$\mathcal{L}(f) = \frac{i}{2\Pi}\int_{\Gamma} f\beta\delta\mathcal{G}_a,$$

$f \in C(\Gamma)$ where β is a complex-valued member of $L(\Gamma)$. This may be concluded from the fact that $M|H_{\mathcal{L}}|$ is quasibounded and so $\mathrm{Re}\, H_{\mathcal{L}}$ and $\mathrm{Im}\, H_{\mathcal{L}}$ are both the difference of quasibounded harmonic functions on Ω. Using (II) we see that β satisfies the condition of Cauchy-Read. Hence $H_{\mathcal{L}}$ is analytic. The sufficiency of (I) and (II) taken together follows.

Continuous boundary function. There remains to be considered the problem of characterizing $A|\Gamma$ for A a continuous map of $\bar{\Omega}$ into X, analytic in Ω. [That A is determined by $A|\Gamma$ is obvious by the maximum principle for subharmonic functions which yields $\max_{\bar{\Omega}}\|A\| = \max_{\Gamma}\|A\|$.] Suppose that $B = A|\Gamma$ for some allowed A. Then we have

$$\text{(III)} \qquad \int_{\Gamma} B\omega = 0 \tag{5.3}$$

for allowed ω as we see from the Cauchy integral theorem.

Suppose that B is a continuous map of Γ into X satisfying (III) for allowed ω. We introduce λ by

$$\lambda(f) = \frac{i}{2\Pi}\int_{\Gamma} fB\delta\mathcal{G}_a,$$

$f \in C(\Gamma)$. The conditions (I) and (II) are fulfilled. $H_\lambda \cup B$ is continuous on $\bar{\Omega}$. Hence $H_\lambda \cup B$ is an allowed A satisfying $A|\Gamma = B$.

To sum up, we have

<u>Theorem</u> 3: <u>Let</u> λ <u>be a continuous linear map of</u> $C(\Gamma)$ <u>into</u> X. <u>A necessary and sufficient condition that</u> $H_\lambda \in \mathbb{H}_p(\Omega, X)$ <u>is that</u> λ <u>satisfy</u> (I) <u>and</u> (II).

<u>Let</u> B <u>be a continuous map of</u> Γ <u>into</u> X. <u>A necessary and sufficient condition that there exist a continuous map</u> A <u>of</u> $\bar{\Omega}$ <u>into</u> X, <u>analytic in</u> Ω, <u>satisfying</u> $A|\Gamma = B$ <u>is that</u> B <u>satisfy</u> (III).

Bibliography

[1] Ahlfors, L., and Sario, L., Riemann Surfaces. Princeton, N. J., 1960.

[2] Behnke, H., and Sommer, F., Theorie der analytischen Funktionen einer komplexen Veränderlichen, 2d. ed. Berlin, 1962.

[3] Behnke, H., and Stein, K., Entwicklung analytischer Funktionen auf Riemannschen Flächen. Math. Ann. 120 (1948) 430-461.

[4] Bonnesen, T., and Fenchel, W., Theorie der konvexen Körper. Berlin, 1934.

[5] Constantinescu, C., and Cornea, A., Ideale Ränder Riemannscher Flächen. Berlin, 1963.

[6] Doob, J., Boundary properties of functions with finite Dirichlet integrals. Ann. Inst. Fourier 12(1962) 573-621.

[7] Doob, J., Remarks on the boundary limits of harmonic functions. J.SIAM Numer. Anal. 3(1966) 229-35.

[8] Duren, P., Theory of H^p Spaces. To appear, Cambridge U. Press.

[9] Earle, C., and Marden, A., On Poincaré series with applications to H^p spaces on bordered Riemann surfaces. Mimeographed.

[10] Gårding,L., and Hörmander, L., Strongly subharmonic functions. Math. Scand. 15 (1964) 93-96. Correction.ibid. 18(1966).

[11] Graves, L., The Theory of Functions of a Real Variable, New York, 1946.

[12] Grenander U., and Szegö, G., Toeplitz Forms and Their Applications, Berkeley/Los Angeles, Cal., 1958.

[13] Hardy, G., The mean value of the modulus of an analytic function. Proc. L. M. S. ser. 2. 14(1915) 269-77.

[14] Heins, M., A lemma on positive harmonic functions. Ann. of Math. 52(1950) 568-73.

[15] Heins, M., Lindelöfian maps. Ann. Math. 62(1955) 418-45.

[16] Heins, M., Selected Topics in the Classical Theory of Functions of a Complex Variable, New York, 1962.

[17] Heins, M.,Symmetric Riemann surfaces and boundary problems. Proc. L. M. S. 3d. ser. 14a(1965) 129-43.

[18] Heins, M., On the theorem of Szegö-Solomentsev Math. Scand. 20(1967) 281-289.

[19] Heins, M., Complex Function Theory. New York, 1968.

[20] Hille, E., and Phillips, R., Functional Analysis and Semi-Groups. Providence, R.I., 1967.

[21] Hoffman, K., Banach Spaces of Analytic Functions. Englewood Cliffs, N. J., 1962.

[22] Kelley, J., Namioka, I., and co-authors, Linear Topological Spaces. Princeton, N. J., 1963.

[23] Martin, R., Minimal positive harmonic functions. Trans. A. M. S. 49(1941) 137-72.

[24] Myrberg, P., Ueber die analytische Fortsetzung von beschränkten Funktionen. Ann. Acad. Sci. Fenn. AI, 58(1949).

[25] Nevanlinna, R., Beschränktartige Potentiale. Math. Nachr. 4(1950-1) 490-501.

[26] Parreau, M., Sur les moyennes des fonctions harmoniques et analytiques et la classification des surfaces de Riemann. Ann. Inst. Fourier 3(1951) 103-97.

[27] Pfluger, A., Theorie der Riemannschen Flächen. Berlin, 1957.

[28] Privalov, I., Boundary Properties of Analytic Functions. Moscow, 1950.(in Russian)

[29] Radó, T., Subharmonic Functions. Berlin, 1937.

[30] Read, A., A converse of Cauchy's theorem and applications to extremal problems. Acta Math. 100(1958) 1-22.

[31] Riesz, F., Oeuvres Complètes, T. 1. Budapest, 1960.

[32] Riesz, M., Sur les fonctions conjuguées. Math.Z. 27(1927) 218-44.

[33] Royden, H., The boundary values of analytic functions. Math.Z. 78(1962) 1-24.

[34] Solomentsev, E., On some classes of subharmonic functions. Bull. Acad. Sci. URSS. Ser. Math. Nr. 516(1938) 571-582.

[35] Stein, P., On a theorem of M. Riesz, J. L. M. S. 8(1933) 242-7.

[36] Szegö, G., Ueber die Randwerte analytischer Funktionen. Math. Ann. 84(1921) 232-44.

[37] Weyl, H., Die Idee der Riemannschen Fläche, 3d. ed. Stuttgart, 1957.

[38] Yamashita, S., On some families of analytic functions on Riemann surfaces. Nagoya Math. J. 31(1968) 57-68.

Offsetdruck: Julius Beltz, Weinheim/Bergstr

Lecture Notes in Mathematics

Bitte wenden / Continued

Beschaffenheit der Manuskripte

Die Manuskripte werden photomechanisch vervielfältigt; sie müssen daher in sauberer Schreibmaschinenschrift geschrieben sein. Handschriftliche Formeln bitte nur mit schwarzer Tusche oder roter Tinte eintragen. Korrekturwünsche werden in der gleichen Maschinenschrift auf einem besonderen Blatt erbeten (Zuordnung der Korrekturen im Text und auf dem Blatt sind durch Bleistiftziffern zu kennzeichnen). Der Verlag sorgt dann für das ordnungsgemäße Tektieren der Korrekturen. Falls das Manuskript oder Teile desselben neu geschrieben werden müssen, ist der Verlag bereit, dem Autor bei Erscheinen seines Bandes einen angemessenen Betrag zu zahlen. Die Autoren erhalten 25 Freiexemplare.

Manuskripte, in englischer, deutscher oder französischer Sprache abgefaßt, nimmt Prof. Dr. A. Dold, Mathematisches Institut der Universität Heidelberg, Tiergartenstraße oder Prof. Dr. B. Eckmann, Eidgenössische Technische Hochschule, Zürich, entgegen.

Cette série a pour but de donner des informations rapides, de niveau élevé, sur des développements récents en mathématiques, aussi bien dans la recherche que dans l'enseignement supérieur. On prévoit de publier

1. des versions préliminaires de travaux originaux et de monographies
2. des cours spéciaux portant sur un domaine nouveau ou sur des aspects nouveaux de domaines classiques
3. des rapports de séminaires
4. des conférences faites à des congrès ou des colloquiums

En outre il est prévu de publier dans cette série, si la demande le justifie, des rapports de séminaires et des cours multicopiés ailleurs et qui sont épuisés.

Dans l'intérêt d'une diffusion rapide, les contributions auront souvent un caractère provisoire; le cas échéant, les démonstrations ne seront données que dans les grandes lignes, et les résultats et méthodes pourront également paraître ailleurs. Par cette série de »prépublications« les éditeurs Springer espèrent rendre d'appréciables services aux instituts de mathématiques par le fait qu'une réserve suffisante d'exemplaires sera toujours disponible et que les personnes intéressées pourront plus facilement être atteintes. Les annonces dans les revues spécialisées, les inscriptions aux catalogues et les copyrights faciliteront pour les bibliothèques mathématiques la tâche de réunir une documentation complète.

Présentation des manuscrits

Les manuscrits, étant reproduits par procédé photomécanique, doivent être soigneusement dactylographiés. Il est demandé d'écrire à l'encre de Chine ou à l'encre rouge les formules non dactylographiées. Des corrections peuvent également être dactylographiées sur une feuille séparée (prière d'indiquer au crayon leur ordre de classement dans le texte et sur la feuille), la maison d'édition se chargeant ensuite de les insérer à leur place dans le texte. S'il s'avère nécessaire d'écrire de nouveau le manuscrit, soit complètement, soit en partie, la maison d'édition se déclare prête à se charger des frais à la parution du volume. Les auteurs recoivent 25 exemplaires gratuits.

Les manuscrits en anglais, allemand ou français peuvent être adressés au Prof. Dr. A. Dold, Mathematisches Institut der Universität Heidelberg, Tiergartenstraße ou au Prof. Dr. B. Eckmann, Eidgenössische Technische Hochschule, Zürich.